Lecture Notes in Mathematics

1491

Editors:
A. Dold, Heidelberg
B. Eckmann, Zürich
F. Takens, Groningen

T0219712

E. Lluis-Puebla   J. L. Loday   H. Gillet
C. Soulé   V. Snaith

Higher Algebraic
K-Theory: an overview

Springer-Verlag

Berlin Heidelberg New York
London Paris Tokyo
Hong Kong Barcelona
Budapest

Authors

Emilio Lluis-Puebla
Departamento de Matématicas
Facultad de Ciencias
Circuito Exterior
Ciudad Universitaria
México 04510 D. F., Mecixo

Jean-Louis Loday
Institut de Recherche Mathématique Avancée
ULP et CNRS, 7, rue René Descartes
67084 Strasbourg, France

Henri Gillet
Department of Mathematics
University of Illinois Chicago
Box 4348
Chicago, IL60680, USA

Christophe Soulé
Institut des Hautes Etudes Scientifiques
35, Route de Chartres
91440 Bures-sur-Yvette, France

Victor Snaith
Department of Mathematics and Statistics
Mc Master University
1280 Main St. W.,
Hamilton, Ontario, Canada L8S 4K1

Mathematics Subject Classification (1980): 19, 11, 13, 14, 55

ISBN 3-540-55007-0 Springer-Verlag Berlin Heidelberg New York
ISBN 0-387-55007-0 Springer-Verlag New York Berlin Heidelberg

© Springer-Verlag Berlin Heidelberg 1992
Printed in Germany

Typesetting: Camera ready by author
Printing and binding: Druckhaus Beltz, Hemsbach/Bergstr.
46/3140-543210 - Printed on acid-free paper

Preface

This volume consists of five different papers on Higher Algebraic K-theory. These are based on several series of lectures delivered during the "First seminar on Algebraic K-theory" at the "Universidad Nacional Autonoma de México" in 1985. Their purpose is to introduce students to this important field by explaining the basic concepts, surveying the main ideas and results, and describing some of the most recent developments.

Each series of lectures has its own perspective (algebra, algebraic geometry or number theory) and is somewhat independent of the other ones. For instance, a basic notion like the cohomology of groups is presented several times. On the other hand, their combination should give a fairly comprehensive overview of the field. Occasionally, results explained in one series of lectures are used (and then explicitly referred to) in another one.

Proofs are not always given, but we hope that the reader will find this volume enjoyable and useful as an introduction to the vast literature.

I would like to thank the Universidad Nacional Autonoma de México and the director of the Faculty of Sciences, Dr. Félix Recillas, for their support and encouragement without which this meeting would not have taken place. Also all my thanks go to Carolina Bello and Ingeborg Jebram for typing this manuscript.

Emilio Lluis-Puebla.

Contents

Introduction to Algebraic K-Theory

Emilio Lluis-Puebla

Departamento de Matemáticas, Facultad de Ciencias, Universidad Nacional Autónoma de México

These notes contain a series of ten lectures delivered at the "Universidad Nacional Autónoma de México" on its "Primer Seminario de K-Teoría Algebraica". They provide an introduction to the subject as well as to the other expositions in this volume.

In Chapter I we review some important concepts from Homological Algebra starting from the elementary concepts and assuming a knowledge of the reader of Group and Ring Theory only. In Chapter II we present the (Co)Homology of Groups in a very elementary way underlying the relevant results used to establish its relation with Algebraic K-Theory.

In Chapter III we define the basic concepts of (classical) Algebraic K-Theory and establish its relation with the Homology of Groups.

Contents

I. Some Homological Algebra

I.1. Modules

Let Λ be a ring (not necessarily commutative) with $1 \neq 0$.

Definition 1.1. A left Λ-*module* or *module over* Λ is a pair (M, μ) where M is an additive abelian group and $\mu : \Lambda \times M \to M$ is a function written $(\alpha, x) \mapsto \alpha x$ such that the following axioms are verified:

i) $\alpha(x + y) = \alpha x + \alpha y$
ii) $(\alpha + \beta)x = \alpha x + \beta x$
iii) $(\alpha\beta)x = \alpha(\beta x)$
iv) $1x = x$ $\qquad (\alpha, \beta \in \Lambda; x, y \in M)$.

μ is called a *scalar multiplication of M* and the elements of Λ are called *scalars*.

For example, take $\Lambda = \mathbb{Z}$, hence any abelian group can be considered as a $\mathbb{Z}$-module. Also if Λ is a field k, a k-module is just a vector space over k. So the concept of Λ-module is a proper generalization of the concepts of abelian group and vector space.

Similarly we can define a right Λ-module using the scalar multiplication (on the right) $\mu : M \times \Lambda \to M$ and writing appropriately the axioms. If Λ is commutative then every left Λ-module is a right Λ-module and vice versa.

This objects (the Λ-modules) are not so special in the sense that (in categorical language) every small abelian category can be considered inside a module category over and adequate ring.

How do we relate two Λ-modules? We relate two sets using functions; we relate two groups using functions that preserve the group structure. So we will relate two modules by means of functions that preserve the Λ-module structure called *homomorphisms*.

Definition 1.2. Let M and N be two Λ-modules. A function $f : M \to N$ is called a Λ-*module homomorphism* if $f(x+y) = f(x) + f(y)$ and $f(\alpha x) = \alpha(f(x))$ for all $\alpha \in \Lambda$; $x, y \in M$.

In Module Theory we also talk about the *kernel* and the *image* of a homomorphism $f : M \to N$ defined as follows:

$$\ker f = \{x \in M | f(x) = 0\}$$
$$\operatorname{im} f = \{f(x) \in N | x \in M\} \ .$$

Also, we define N to be a *submodule* of a Λ-module M if N is a subgroup of M and for all $\alpha \in \Lambda$, $\alpha N = \{\alpha x | x \in N\} \subset N$.

The composition of homomorphisms turns out to be a homomorphism; the image under a homomorphism of a submodule is a submodule; the inverse image of a submodule under a homomorphism is a submodule; and, in particular the kernel and the image of a homomorphism are submodules.

We also have the concept of quotient module whose elements are the distinct cosets of a submodule in a module. In particular we define the *cokernel* of a module homomorphism $f : M \to N$ as the quotient module: coker $f = N/\operatorname{im} f$.

We will relate several Λ-modules by means of a collection of homomorphisms as follows:

Definition 1.3. We will say that a sequence of Λ-modules

$$\ldots \to M_{i-1} \xrightarrow{f_{i-1}} M_i \xrightarrow{f_i} M_{i+1} \xrightarrow{f_{i+1}} \ldots$$

is *semiexact* in M_i if $\operatorname{im} f_{i-1} \subset \ker f_i$. If it is semiexact in each module we will call it a *semiexact sequence*.

This definition is equivalent to say that the composition $f \circ f_{i-1}$ is the *trivial homomorphism* denoted by 0, i.e. $f_i \circ f_{i-1} = 0$. We will say that the sequence of (1.3) is *exact* in M_i if it is semiexact and $\operatorname{im} f_{i-1} \supset \ker f_i$, i.e. if $\operatorname{im} f_{i-1} = \ker f_i$. If it is exact in each M_i it is called an *exact sequence*. Of course, every exact sequence is a semiexact sequence but the converse is not true.

An exact sequence of the form

$$0 \to M' \xrightarrow{f} M \xrightarrow{g} M'' \to 0$$

where 0 denotes the trivial module is called a *short exact sequence*. It is immediate that f is an injective homomorphism called *monomorphism* and g is an onto homomorphism called *epimorphism*. The following notation is used to represent a short exact sequence:

$$M' \xrightarrow{f} M \xrightarrow{g} M'' \ .$$

It is just a mask of a submodule and a quotient module of a Λ-module M in the exact sequence

$$N \rightarrowtail M \twoheadrightarrow M/N \ .$$

There is a category Mod_Λ whose objects are the Λ-modules and whose morphisms are the homomorphisms of Λ-modules. Mod_Λ is studied by analyzing the behavior of certain functors defined on it. The most important are hom and $\otimes$ and certain functors derived from them.

Let $\hom_\Lambda(M, N)$ denote the set of all homomorphisms from the Λ-module M to the Λ-module N. If Λ is not commutative then $\hom_\Lambda(M, N)$ is always an abelian group and if Λ is commutative $\hom_\Lambda(M, N)$ is a Λ-module.

The following theorem can be considered as the beginning of Homological Algebra.

Theorem 1.4. *a) If $N' \xrightarrow{\psi} N \xrightarrow{\psi'} N''$ is an exact sequence of Λ-modules then there is an exact induced sequence*

$$0 \longrightarrow \hom_\Lambda(M, N') \xrightarrow{\psi_*} \hom_\Lambda(M, N) \xrightarrow{\psi'_*} \hom_\Lambda(M, N'') \ .$$

b) If $M'\xrightarrow{\varphi} M\xrightarrow{\varphi'} M''$ is an exact sequence of Λ-modules then there is an exact induced sequence

$$\hom_\Lambda(M',N)\xleftarrow{\varphi^*}\hom_\Lambda(M,N)\xleftarrow{\varphi'^*}\hom_\Lambda(M'',N)\longleftarrow 0 \ .$$

We can expect that the case when ψ'_* is onto be of interest. It really is. The class of Λ-modules that satisfy ψ'_* to be onto whenever ψ' is onto are called *projective* Λ-modules. By imitating the case of k-modules (k a field) i.e. vector spaces, we say that a Λ-module is *free* if it has a *basis*. It turns out that every free module is projective and that every Λ-module is a quotient of a free module. Dually we can define the concept of an *injective* Λ-module as the one who makes φ^* an epimorphism whenever φ is a monomorphism. It turns out that every module is isomorphic to a submodule of an injective module, being this a very important fact.

For M a right Λ-module and N a left Λ-module we define an abelian group $M\otimes_\Lambda N$, called the *tensor product* of M and N over Λ, as the free abelian group generated by the symbols $x\otimes y$, $x\in M$, $y\in N$ modulo the subgroup generated by the expressions

$$(x+x')\otimes y - (x\otimes y + x'\otimes y)$$
$$x\otimes(y+y') - (x\otimes y + x\otimes y')$$
$$x\lambda\otimes y - x\lambda\otimes y \qquad \lambda\in\Lambda,\ x,x'\in M;\ y,y'\in N \ .$$

If Λ is commutative $M\otimes_\Lambda N$ is a Λ-module.

We have a theorem similar to 1.4:

Theorem 1.5. *a) If $N'\xrightarrow{\psi} N\xrightarrow{\psi'} N''$ is an exact sequence of Λ-modules then there is an exact induced sequence*

$$M\otimes N'\xrightarrow{1_M\otimes\psi} M\otimes_\Lambda N'\xrightarrow{1_N\otimes\psi'} M\otimes_\Lambda N''\longrightarrow 0 \ .$$

b) If $M'\xrightarrow{\varphi} M\xrightarrow{\varphi'} M''$ is an exact sequence of Λ-modules then there is an exact induced sequence

$$M'\otimes_\Lambda N\xrightarrow{\varphi\otimes 1_N} M\otimes_\Lambda N\xrightarrow{\varphi'\otimes 1_N} M''\otimes_\Lambda N\longrightarrow 0 \ .$$

Observe that $(\hom_{\Lambda-},\_)$ is a functor of two variables from the category of Λ-modules Mod_Λ, to the category of abelian groups Ab. It is covariant on the second variable and contravariant on the first one.

Also $\_\otimes_\Lambda\_$ is a functor of two variables from Mod_Λ to Ab and is covariant on both variables.

One extremely important relation that ties the functors hom and $\otimes$ is given by the following isomorphism

$$\hom_{\Lambda'}(M\otimes_\Lambda N, U) \cong \hom_\Lambda(M, \hom_{\Lambda'}(N, U))$$

where M and N are Λ-modules (right and left respectively) and N and U are Λ'-modules (both right).

Exercise 1.6. Prove that if

$$0 \to M' \to M \to M'' \to 0$$

is a short exact sequence that splits then so are

$$0 \to M' \otimes_\Lambda N \to M \otimes_\Lambda N \to M'' \otimes_\Lambda N \to 0$$

and

$$0 \to \hom_\Lambda(M'', N) \to \hom_\Lambda(M, N) \to \hom_\Lambda(M', N) \to 0 \ .$$

I.2. Resolutions and Homology

Let $\{C_n\}_{n \in \mathbb{Z}}$ be a family of Λ-modules and $\{\partial_n : C_n \to C_{n-1}\}_{n \in \mathbb{Z}}$ a family of Λ-module homomorphisms such that $\partial_n \circ \partial_{n+1} = 0$. A *chain complex* or *chain* over Λ is the pair $C = \{C_n, \partial_n)$ and we write it as follows:

$$C : \ldots \xrightarrow{\partial_{n+2}} C_{n+1} \xrightarrow{\partial_{n+1}} C_n \xrightarrow{\partial_n} C_{n-1} \xrightarrow{\partial_{n-1}} \ldots \ .$$

This means that a chain is just a decreasing semiexact sequence of Λ-modules. A morphism $\varphi : C \to D$ between two chains is a family of Λ-module homomorphisms $\{\varphi_n : C_n \to D_n\}$ such that the following diagram commutes:

$$
\begin{array}{ccccccccc}
C: & \cdots & \longrightarrow & C_n & \xrightarrow{\partial_n} & C_{n-1} & \xrightarrow{\partial_{n-1}} & C_{n-2} & \longrightarrow & \cdots \\
& \downarrow{\varphi} & & \downarrow{\varphi_n} & & \downarrow{\varphi_{n-1}} & & \downarrow{\varphi_{n-2}} & & \\
D: & \cdots & \longrightarrow & D_n & \xrightarrow{\partial'_n} & D_{n-1} & \xrightarrow{\partial'_{n-1}} & D_{n-2} & \longrightarrow & \cdots \ .
\end{array}
$$

The main concept in Homological Algebra is the following: Let C be a chain complex. Then the n-th homology module of C, denoted $H_n(C)$, is the quotient module $H_n(C) = \ker \partial_n / \operatorname{im} \partial_{n+1}$.

$H_n(C)$ measures the inexactness of the chain C. For example, if C is exact then $\operatorname{im} \partial_{n+1} = \ker \partial_n$, hence $H_n(C) = 0$. We have associated to a chain C a graded module $H_*(C) = \{H_n(C)\}$ which we call *the homology of the chain C*. A chain morphism induces a well defined morphism (of degree 0) $\varphi_* : H_*(C) \to H_*(D)$ between graded modules. Then $H_*(\_)$ is a covariant functor from the category of chain complexes to the category of graded Λ-modules.

If we consider semiexact families $\{C^n\}_{n \in \mathbb{Z}}$ with increasing index we obtain dual concepts; we have cochains, cochain morphisms, cohomology of a cochain, etc.

Given two chain complexes C, D and two morphisms between them $\varphi, \varphi' : C \to D$ when do they induce the same homomorphism between $H_*(C)$ and $H_*(D)$?

To answer this question we introduce the concept of *homotopy*.

Definition 2.1. Let $C = \{C_n, \partial_n\}$ and $D = \{D_n, \partial'_n\}$ be two chain complexes and $\varphi, \varphi' : C \to D$ two chain morphisms. We will say that φ is *homotopic* to φ' if there exists a family of Λ-module homomorphisms

$$h = \{h_n : C_n \to D_{n+1} \mid n \in \mathbb{Z}\} \ .$$

such that $\partial'_{n+1} \circ h_n + h_{n-1} \circ \partial_n = \varphi_n - \varphi'_n$ for all $n \in \mathbb{Z}$ in the following diagram

$$
\begin{array}{ccccccccc}
C: & \cdots & \longrightarrow & C_{n+1} & \xrightarrow{\partial_{n+1}} & C_n & \xrightarrow{\partial_n} & C_{n-1} & \xrightarrow{\partial_{n-1}} & \cdots \\
& & & \varphi'_{n+1}\downarrow\downarrow\varphi_{n+1} & {}^{h_n}\swarrow & \varphi'_n\downarrow\downarrow\varphi_n \ {}^{h_{n-1}}\swarrow & & \varphi'_{n-1}\downarrow\downarrow\varphi_{n-1} & & \\
D: & \cdots & \longrightarrow & D_{n+1} & \xrightarrow[\partial'_{n+1}]{} & D_n & \xrightarrow[\partial'_n]{} & D_{n-1} & \xrightarrow[\partial'_{n-1}]{} & \cdots
\end{array}
$$

The family $h = \{h_n\}$ is called a *chain homotopy* and we will say that φ *is homotopic to* φ'. In symbols

$$h : \varphi \sim \varphi' : C \to D \ .$$

It is easy to check that $\sim$ is an equivalence relation.

We will say that a chain morphism $\varphi : C \to D$ is a *homotopy equivalence* if there is a chain morphism $\varphi' : D \to C$ such that $\varphi' \circ \varphi \sim 1_C$ and $\varphi \circ \varphi' \sim 1_D$. In this case we will say that C and D are of the same *homotopy type*.

Theorem 2.2. *If* $\varphi \sim \varphi' : C \to D$ *then* $H_*(\varphi) = H_*(\varphi') : H_*(C) \to H_*(D)$.

Proof. Let $h : \varphi \sim \varphi'$ be the homotopy. Let $x \in H_n(C)$ be arbitrary, let $z \in Z_n(C)$ such that $p(z) = x$ where $p : Z_n(C) \to H_n(C)$ is the projection. Then

$$\varphi_n(z) - \varphi'_n(z) = \partial'_{n+1}h(z) + h_{n-1}\partial_n(z) = \partial'_{n+1}h_n(z)$$

because $\partial_n(z) = 0$. Since $\partial'_{n+1}h_n(z) \in B_n(D)$,

$$[H_*(\varphi)](x) = [H_*(\varphi')](x) \ .$$

Then $H_n(\varphi) = H_n(\varphi')$ for all $n \in \mathbb{Z}$, i.e. $\varphi(z)$ and $\varphi'(z)$ are homologous. $\square$

The converse of this theorem is not true.

If $\varphi = 0 : C \to C$ is the trivial morphism and $\varphi' = 1_C : C \to C$ is the identity morphism then a homotopy $h : \varphi \sim \varphi'$ is called a *contraction* and we have $\partial'_{n+1} \circ h_n + h_{n-1} \circ \partial_n = 1$, $n \in \mathbb{Z}$. This implies by theorem 2.2 that $H_*(C) = 0$ and that C is exact.

Since the category of chain complexes is an abelian category we can form short exact sequences of chain complexes displayed vertically as follows:

$$
\begin{array}{ccccccccc}
0 & \longrightarrow & C & \xrightarrow{\varphi} & D & \xrightarrow{\varphi'} & E & \longrightarrow & 0 \\
 & & \vdots & & \vdots & & \vdots & & \\
 & & \downarrow & & \downarrow & & \downarrow & & \\
0 & \longrightarrow & C_{n+1} & \xrightarrow{\varphi_{n+1}} & D_{n+1} & \xrightarrow{\varphi'_{n+1}} & E_{n+1} & \longrightarrow & 0 \\
 & & \downarrow{\partial_{n+1}} & & \downarrow{\partial'_{n+1}} & & \downarrow{\partial''_{n+1}} & & \\
0 & \longrightarrow & C_n & \xrightarrow{\varphi_n} & D_n & \xrightarrow{\varphi'_n} & E_n & \longrightarrow & 0 \\
 & & \downarrow{\partial_n} & & \downarrow{\partial'_n} & & \downarrow{\partial''_n} & & \\
0 & \longrightarrow & C_{n-1} & \xrightarrow{\varphi_{n-1}} & D_{n-1} & \xrightarrow{\varphi'_{n-1}} & E_{n-1} & \longrightarrow & 0 \\
 & & \downarrow & & \downarrow & & \downarrow & & \\
 & & \vdots & & \vdots & & \vdots & &
\end{array}
$$

We have the following basic theorem:

Theorem 2.3. *Let $C \rightarrowtail D \twoheadrightarrow E$ be a short exact sequence of chain complexes. Then there is a homomorphism $\kappa_n : H_n(E) \to H_{n-1}(C)$ for each $n \in \mathbb{Z}$ such that the following sequence is exact:*

$$\ldots \longrightarrow H_n(C) \longrightarrow H_n(D) \longrightarrow H_n(E) \xrightarrow{\kappa_n}$$
$$\xrightarrow{\kappa_n} H_{n-1}(C) \longrightarrow H_{n-1}(D) \longrightarrow H_{n-1}(E) \xrightarrow{\kappa_{n-1}} \ldots \ .$$

Consider a positive exact chain complex of projective (free) Λ-modules $P = \{P_n, \partial_n\}$, that is, such that $H_n(P) = 0$ for $n \geq 1$ and let us assume it satisfies that $H_0(P) \cong M$. We will write it as follows

$$P : \ldots \longrightarrow P_n \xrightarrow{\partial_n} P_{n-1} \xrightarrow{\partial_{n-1}} \ldots \longrightarrow P_1 \xrightarrow{\partial_1} P_0 \xrightarrow{\varepsilon} M \longrightarrow 0$$

and call it a *projective (free) resolution* of a Λ-module M.

Proposition 2.4. *Let M be a Λ-module. Then there exists a projective resolution of M.*

Proof. Since every Λ-module is quotient of a free Λ-module there is a short exact sequence

$$0 \longrightarrow M_0 \xrightarrow{\mu_0} F_0 \xrightarrow{\eta_0} M \longrightarrow 0$$

where F_0 is a free Λ-module. Since M_0 is a quotient of a free Λ-module F_1, there is a short exact sequence

$$0 \longrightarrow M_1 \xrightarrow{\mu_1} F_1 \xrightarrow{\eta_1} M_0 \longrightarrow 0$$

where F_1 is free. By induction we obtain a short exact sequence

$$0 \longrightarrow M_n \xrightarrow{\mu_n} F_n \xrightarrow{\eta_n} M_{n-1} \longrightarrow 0$$

with F_n free. Define a sequence

$$F : \ldots \longrightarrow F_{n+1} \xrightarrow{\partial_{n+1}} F_n \xrightarrow{\partial_n} F_{n-1} \longrightarrow \ldots$$

by

$$F_n = \begin{cases} M & \text{if } n = -1 \\ F_n & \text{if } n \geq 0 \\ 0 & \text{if } n < -1 \end{cases} \qquad \partial_n = \begin{cases} \eta_0 & \text{if } n = 0 \\ \mu_{n-1} \circ \eta_n & \text{if } n \geq 0 \\ 0 & \text{if } n < 0 \end{cases}$$

Since μ_n is a monomorphism and η_n is an epimorphism we have that

$$\text{im } \partial_{m+1} = \text{im } \mu_n = \ker \eta_n = \ker \partial_n .$$

Then the sequence is exact and clearly $H_0(F) \cong M$. Since every free module is projective we are done. $\square$

Definition 2.5. A positive chain complex $C = \{C_n, \partial_n\}$ is called *acyclic* if $H_n(C) = 0$ for $n \geq 1$ (i.e. C is exact till C_1, $H_0(C)$ may be different of 0). Equivalently, the seqence

$$\ldots \longrightarrow C_n \longrightarrow C_{n-1} \longrightarrow \ldots \longrightarrow C_1 \xrightarrow{\partial_1} C_0 \xrightarrow{\varepsilon} H_0(C) \longrightarrow 0$$

is exact.

We can say, in view of 2.5, that a projective resolution of a Λ-module M is a projective and acyclic chain $P = \{P_n, \partial_n\}$ such that $H_0(P) \cong M$.

The following lemma is considered as the fundamental lemma in Homological Algebra. It tells us how we can construct chain morphisms and homotopies from a projective chain into an acyclic chain.

Lemma 2.6. *Let* $C = \{C_n, \partial_n\}$ *and* $D = \{D_n, \partial'_n\}$ *be two chain complexes. Let* $\varphi = \{\varphi_i : C_i \to D_i\}_{i \leq n}$ *be a family of* Λ-*module homomorphisms such that* $\partial'_i \circ \varphi_i = \varphi_{i-1} \circ \partial_i$ *for* $i \leq n$. *Suppose that* C_i *is projective for* $i > n$ *and that* $H_i(D) = 0$ *for* $i \geq n$. *Then* $\{\varphi_i\}_{i \leq n}$ *extends to a chain morphism* $\varphi : C \to D$ *and is unique up to homotopy.*

Definition 2.7. Let P be a projective resolution of a Λ-module M

$$P : \ldots \longrightarrow P_n \longrightarrow P_{n-1} \longrightarrow \ldots \longrightarrow P_1 \longrightarrow P_0 \xrightarrow{\varepsilon} M \longrightarrow 0 .$$

A *reduced projective resolution* P_M of M is a projective resolution of M in which M has been suppressed, i.e.

$$P_M : \ldots \longrightarrow P_n \longrightarrow P_{n-1} \longrightarrow \ldots \longrightarrow P_1 \xrightarrow{\partial_1} P_0 \longrightarrow 0 \ .$$

Observe that we have not lost any information about P since $M = \operatorname{coker} \partial_1$. We consider the projective resolutions as a generalization of a Λ-module presentation, i.e. a generalization of the concepts of generators and relations. The advantage of P_M is that it consists only of projective Λ-modules.

Now, lets compare two projective resolutions of a Λ-module.

Theorem 2.8. *Let P and P' be projective resolutions of a Λ-module M. Then there exist a chain morphism $\varphi : P \to P'$ such that $\varepsilon' \varphi_0 = \varepsilon$. Furthermore, φ is unique up to homotopy and is a homotopy equivalence.*

Proof. By lemma 2.6 applied to $n = -1$ we obtain a chain morphism $\varphi : P \to P'$ such that $\varepsilon' \varphi_0 = \varepsilon$. Furthermore, φ is unique up to homotopy ($h_{-1} = 0$). Similarly there exists $\varphi' : P' \to P$. By 2.6, the composition $\varphi' \varphi : P \to P$ and the identity $1_P : P \to P$ are homotopic, i.e. $\varphi' \varphi \sim 1_P$. Analogously $\varphi \varphi' \sim 1_{P'}$. Then φ is a homotopy equivalence. □

By the previous theorem we can say that two projective resolutions of a Λ-module M are of the *same homotopy type* or that they are *unique up to homotopy equivalence*.

Example 2.9. Consider $\mathbb{Z}$-modules. The subgroups of a free group are free. Then, any abelian group G admits a free resolution of length ≤ 1:

$$0 \longrightarrow L_1 \longrightarrow L_0 \longrightarrow G \longrightarrow 0 \ .$$

For example, the $\mathbb{Z}$-module $\mathbb{Z}/p$, p a prime, admits the following resolution

$$0 \longrightarrow \mathbb{Z} \xrightarrow{\mu} \mathbb{Z} \longrightarrow \mathbb{Z}/p \longrightarrow 0$$

where μ is multiplication by p.

I.3. Torsion and Extension Functors

Let $P_M : \ldots \longrightarrow P_n \xrightarrow{\partial_n} P_{n-1} \xrightarrow{\partial_{n-1}} \ldots \longrightarrow P_1 \xrightarrow{\partial_1} P_0 \longrightarrow 0$ be a reduced projective resolution of a right Λ-module M. Let N be a left Λ-module and consider the tensor product $P_M \otimes_\Lambda N$ which is the sequence

$$P_M \otimes_\Lambda N : \ldots \longrightarrow P_n \otimes_\Lambda N \xrightarrow{\partial_n \otimes 1} P_{n-1} \otimes_\Lambda N \xrightarrow{\partial_{n-1} \otimes 1}$$

$$\ldots \longrightarrow P_1 \otimes_\Lambda N \xrightarrow{\partial_1 \otimes 1} P_0 \otimes_\Lambda N \longrightarrow 0 \ .$$

$P_M \otimes_\Lambda N$ is a semiexact sequence since, for all $n > 1$,

$$(\partial_{n-1} \otimes_\Lambda 1) \circ (\partial_n \otimes_\Lambda 1) = (\partial_{n-1} \circ \partial_n) \otimes_\Lambda 1 = 0 \otimes_\Lambda 1 = 0 \ .$$

Then we can form

$$H_* \left(P_M \otimes_\Lambda N \right) = \{ H_n \left(P_M \otimes_\Lambda N \right) \}_{n \geq 0}$$

and have the following

Definition 3.1. For each $n \geq 0$, let $\mathrm{Tor}_n^\Lambda(M, N)$ denote $H_n(P_M \otimes_\Lambda N)$ and call it the n-th *Tor group over Λ of M and N*.

It is easy to see that $\mathrm{Tor}_n^\Lambda(M, N)$ is independent of the choice of the resolution. It only depends on n, M and N.

Let $f : M \to M''$ and $g : N \to N''$ be Λ-module homomorphisms. Let P_M and $P_{M''}$ be reduced projective resolutions of M and M'' respectively. By lemma 2.6 there is a chain morphism $\varphi : P_M \to P_{M''}$ that extends f. Then $\varphi \otimes g$ is a chain morphism that induce

$$(\varphi \otimes g)_* : H_* \left(P_M \otimes_\Lambda N \right) \longrightarrow H_* \left(P_M \otimes_\Lambda N'' \right)$$

i.e.

$$(\varphi \otimes g)_* : \mathrm{Tor}_*^\Lambda(M, N) \longrightarrow (M'', N'') \ .$$

It does not depend on φ but only on n, f and g. It is easy to prove the following

Proposition 3.2. $\mathrm{Tor}_n^\Lambda(\_, \_)$ *is a covariant bifunctor from the category of Λ-modules to the category of abelian groups.*

We leave to the reader the proof of the following

Exercise . Let $M' \rightarrowtail M \twoheadrightarrow M''$ be a short exact sequence of Λ-modules. Let $P_{M'}'$ and $P_{M''}''$ be reduced projective resolutions of M' and M'' respectively. Then there is a reduced projective resolution P_M of M such that

$$P_{M'}' \rightarrowtail P_M \twoheadrightarrow P_{M''}''$$

is an exact sequence of reduced projective resolutions that splits.

Let $N' \rightarrowtail N \twoheadrightarrow N''$ be a short exact sequence of left Λ-modules and P_M a reduced projective resolution of a right Λ-module M. Then

$$P_M \otimes_\Lambda N' \rightarrowtail P_M \otimes_\Lambda N \twoheadrightarrow P_M \otimes_\Lambda N''$$

is a short exact sequence of chains. By theorem 2.3 there is a homomorphism

$$\kappa_n : H_n \left(P_M \otimes_\Lambda N'' \right) \longrightarrow H_{n-1} \left(P_M \otimes_\Lambda N' \right)$$

such that the following sequence is exact:

$$\ldots \longrightarrow H_n \left(P_M \otimes_\Lambda N' \right) \longrightarrow H_n \left(P_M \otimes_\Lambda N \right) \longrightarrow H_n \left(P_M \otimes_\Lambda N'' \right)$$
$$\xrightarrow{\kappa_n} H_{n-1} \left(P_M \otimes_\Lambda N' \right) \longrightarrow H_{n-1} \left(P_M \otimes_\Lambda N \right) \longrightarrow H_{n-1} \left(P_M \otimes_\Lambda N'' \right) \longrightarrow \ldots \ .$$

Then using theorem 3.2 we have

Theorem 3.3. *Let $N' \rightarrowtail N \twoheadrightarrow N''$ be an exact sequence of Λ-modules and M a Λ-module. Then there exists a long exact sequence*

$$\ldots \longrightarrow \operatorname{Tor}_n^\Lambda(M, N') \longrightarrow \operatorname{Tor}_n^\Lambda(M, N) \longrightarrow \operatorname{Tor}_n^\Lambda(M, N'')$$
$$\longrightarrow \operatorname{Tor}_{n-1}^\Lambda(M, N') \longrightarrow \ldots \longrightarrow \operatorname{Tor}_0^\Lambda(M, N'') \longrightarrow 0 .$$

A companion theorem for 3.3 is

Theorem 3.4. *Let $M' \rightarrowtail M \twoheadrightarrow M''$ be a short exact sequence of Λ-modules and N a Λ-module. Then there exists a long exact sequence*

$$\ldots \longrightarrow \operatorname{Tor}_n^\Lambda(M', N) \longrightarrow \operatorname{Tor}_n^\Lambda(M, N) \longrightarrow \operatorname{Tor}_n^\Lambda(M'', N)$$
$$\longrightarrow \operatorname{Tor}_{n-1}^\Lambda(M', N) \longrightarrow \ldots \longrightarrow \operatorname{Tor}_0^\Lambda(M'', N) \longrightarrow 0 .$$

Proof. Use the previous exercise to obtain a short exact sequence of chains that split

$$P'_{M'} \rightarrowtail P_M \twoheadrightarrow P''_{M''} .$$

Then by exercise 1.6,

$$P'_{M'} \otimes_\Lambda N \rightarrowtail P_M \otimes_\Lambda N \twoheadrightarrow P''_{M''} \otimes_\Lambda N$$

is a short exact sequence of chains that split. Finally apply theorem 2.3. $\square$

We can also define the torsion functor as follows: Let Q_N be a reduced projective resolution of N. Consider the tensor product $M \otimes_\Lambda Q_N$ which is a semiexact sequence whose n-th homology $H_n(M \otimes_\Lambda Q_N)$ will be denoted by $\overline{\operatorname{Tor}}_n^\Lambda(M, N)$.

Before we prove that the definitions of Tor and $\overline{\operatorname{Tor}}$ coincide lets mention some important results for $\operatorname{Tor}_n^\Lambda(\_, \_)$ and $\overline{\operatorname{Tor}}_n^\Lambda(\_, \_)$.

Theorem 3.5. *a) The functors $\operatorname{Tor}_0^\Lambda(\_, N)$ and $\overline{\operatorname{Tor}}_0^\Lambda(M, \_)$ are naturally equivalent to the functors $\_ \otimes_\Lambda N$ and $M \otimes_\Lambda \_$ respectively.*
b) Let P be a projective Λ-module. Then

$$\operatorname{Tor}_n^\Lambda(P, N) = 0 = \overline{\operatorname{Tor}}_n^\Lambda(M, P)$$

for any Λ-modules M, N and $n \geq 1$.
c) If P is a flat Λ-module (i.e. P is a flat Λ-module if given a short exact sequence of Λ-modules $M' \rightarrowtail M \twoheadrightarrow M''$ then the induced sequence $M' \otimes_\Lambda P \longrightarrow M \otimes_\Lambda P \longrightarrow M'' \otimes_\Lambda P$ is short exact) then $\operatorname{Tor}_n^\Lambda(P, N) = 0$ for all $n \geq 1$.
d) If $\operatorname{Tor}_1^\Lambda(P, N) = 0$ for all N then P is a flat Λ-module.

The following theorem enables us to compute the torsion product either via a reduced projective resolution of the first variable or of the second one.

Theorem 3.6. $\operatorname{Tor}_n^\Lambda(M, N) \cong \overline{\operatorname{Tor}}_n^\Lambda(M, N)$.

Proof. Let $K \rightarrowtail P \twoheadrightarrow M$ be a projective presentation of M, i.e. a short exact sequence with P projective. By 3.4 we have a long exact sequence

$$\ldots \longrightarrow \operatorname{Tor}_n^\Lambda(K, N) \longrightarrow \operatorname{Tor}_n^\Lambda(P, N) \longrightarrow \operatorname{Tor}_n^\Lambda(M, N) \longrightarrow \ldots$$
$$\ldots \longrightarrow K \otimes_\Lambda N \longrightarrow P \otimes_\Lambda N \longrightarrow M \otimes_\Lambda N \longrightarrow 0 .$$

By a corresponding theorem for $\overline{\operatorname{Tor}}_n^\Lambda$ analogous to 3.3 we have a long exact sequence

$$\ldots \longrightarrow \overline{\operatorname{Tor}}_n^\Lambda(K, N) \longrightarrow \overline{\operatorname{Tor}}_n^\Lambda(P, N) \longrightarrow \overline{\operatorname{Tor}}_n^\Lambda(M, N) \longrightarrow \ldots$$
$$\ldots \longrightarrow K \otimes_\Lambda N \longrightarrow P \otimes_\Lambda N \longrightarrow M \otimes_\Lambda N \longrightarrow 0 .$$

Using 3.5(b) such long exact sequences becomes

$$0 \longrightarrow \operatorname{Tor}_1^\Lambda(M, N) \longrightarrow K \otimes_\Lambda N \longrightarrow P \otimes_\Lambda N$$
$$0 \longrightarrow \overline{\operatorname{Tor}}_1^\Lambda(M, N) \longrightarrow K \otimes_\Lambda N \longrightarrow P \otimes_\Lambda N \qquad (*)$$

for $n = 1$ and

$$0 \longrightarrow \operatorname{Tor}_n^\Lambda(M, N) \longrightarrow \operatorname{Tor}_{n-1}^\Lambda(K, N) \longrightarrow 0$$
$$0 \longrightarrow \overline{\operatorname{Tor}}_n^\Lambda(M, N) \longrightarrow \overline{\operatorname{Tor}}_{n-1}^\Lambda(K, N) \longrightarrow 0 \qquad (**)$$

for $n \geq 2$.

Clearly $\operatorname{Tor}_0^\Lambda(M, N) = \overline{\operatorname{Tor}}_0^\Lambda(M, N) = M \otimes_\Lambda N$. By the sequences of $(*)$ the theorem is valid for $n = 1$ and by $(**)$ using induction over n the theorem is proved. $\qquad \square$

Lets define another fundamental functor in Homological Algebra. Let

$$P_M : \ldots \longrightarrow P_n \xrightarrow{\partial_n} P_{n-1} \xrightarrow{\partial_{n-1}} \ldots \longrightarrow P_1 \longrightarrow P_0 \longrightarrow 0$$

be a reduced projective resolution of a Λ-module M. Let N be a Λ-module and consider $\hom_\Lambda(P_M, N)$ which is the sequence

$$\hom_\Lambda(P_M, N) : \ldots \longleftarrow \hom_\Lambda(P_n, N) \overset{\hom_\Lambda(\partial_n, 1)}{\longleftarrow} \ldots$$
$$\longleftarrow \hom_\Lambda(P_1, N) \longleftarrow \hom_\Lambda(P_0, N) \longleftarrow 0 .$$

$\hom_\Lambda(P_M, N)$ is a semiexact sequence since, for all $n > 1$

$$\hom_\Lambda(\partial_n, 1) \circ \hom_\Lambda(\partial_{n-1}, 1) = \hom_\Lambda(\partial_{n-1} \circ \partial_n, 1) = \hom_\Lambda(0, 1) = 0 .$$

Then we can form

$$H^*(\hom_\Lambda(P_M, N)) = \{H^n(\hom_\Lambda(P_M, N))\}_{n \geq 0}$$

and we have the following

Definition 3.7. For each $n \geq 0$, let $\operatorname{Ext}_\Lambda^n(M, N)$ denote $H^n(\hom_\Lambda(P_M, N))$ and call it the n-th *Ext group over Λ of M and N.*

$\operatorname{Ext}_\Lambda^n(M, N)$ is independent of the choice of the resolution.

Let $f : M \to M''$ and $g : N \to N''$ be Λ-module homomorphisms. Let P_M and P_M'' be reduced projective resolutions of M and M'' respectively. By lemma 2.6 there is a chain morphism $\varphi : P_M \to P_{M''}''$ that extends f. Then $\hom_\Lambda(\varphi, g)$ is a cochain morphism that induces a homomorphism

$$\hom_\Lambda(\varphi, g)^* : H^*(\hom(P_{M''}, N)) \longrightarrow H^*(\hom(P_M, N''))$$

i.e.

$$\hom_\Lambda(\varphi, g)^* : \operatorname{Ext}_\Lambda^n(M'', N) \longrightarrow \operatorname{Ext}_\Lambda^n(M, N'') \ .$$

It does not depend on φ but only on n, f and g. Then we have the following

Theorem 3.8. $\operatorname{Ext}_\Lambda^n(\_,\_)$ *is a bifunctor from the category of Λ-modules to the category of abelian groups. It is contravariant in the first variable and covariant in the second variable.*

Let $N' \rightarrowtail N \twoheadrightarrow N''$ be a short exact sequence of Λ-modules and P_M a reduced projective resolution of M. Then

$$\hom_\Lambda(P_M, N') \rightarrowtail \hom_\Lambda(P_M, N) \twoheadrightarrow \hom_\Lambda(P_M, N'')$$

is a short exact sequence of cochains. By the dual of theorem 2.3 there is a homomorphism

$$\kappa^n : H^n(\hom_\Lambda(P_M, N'')) \longrightarrow H^{n+1}(\hom_\Lambda(P_M, N'))$$

such that the following sequence is exact:

$$\ldots \longrightarrow H^n(\hom_\Lambda(P_M, N')) \longrightarrow H^n(\hom_\Lambda(P_M, N)) \longrightarrow H^n(\hom_\Lambda(P_M, N''))$$
$$\xrightarrow{\kappa^n} H^{n+1}(\hom_\Lambda(P_M, N')) \longrightarrow \ldots \ .$$

Using theorem 3.8 and the previous one we have the following:

Theorem 3.9. *Let $N' \rightarrowtail N \twoheadrightarrow N''$ be a short exact sequence of Λ-modules and M a Λ-module. Then there is a long exact sequence*

$$0 \longrightarrow \operatorname{Ext}_\Lambda^0(M, N') \longrightarrow \ldots \longrightarrow \operatorname{Ext}_\Lambda^n(M, N'') \xrightarrow{\kappa^n}$$
$$\xrightarrow{\kappa^n} \operatorname{Ext}_\Lambda^{n+1}(M, N') \longrightarrow \operatorname{Ext}_\Lambda^{n+1}(M, N) \longrightarrow \operatorname{Ext}_\Lambda^{n+1}(M, N'') \longrightarrow \ldots \ .$$

$\square$

The corresponding theorem for 3.9 is

Theorem 3.10. *Let* $M' \rightarrowtail M \twoheadrightarrow M''$ *be a short exact sequence of* Λ-*modules and* N *a* Λ-*module. Then there is a long exact sequence*

$$0 \longrightarrow \operatorname{Ext}_\Lambda^0(M'', N) \longrightarrow \ldots \longrightarrow \operatorname{Ext}_\Lambda^n(M', N)$$
$$\longrightarrow \operatorname{Ext}_\Lambda^{n+1}(M'', N) \longrightarrow \operatorname{Ext}_\Lambda^{n+1}(M, N) \longrightarrow \operatorname{Ext}_\Lambda^{n+1}(M', N) \longrightarrow \ldots .$$

Proof. Use lemma 3.4 and exercise 1.6 to obtain a short exact sequence of cochains

$$0 \longrightarrow \hom_\Lambda(P_{M''}'', N) \longrightarrow \hom_\Lambda(P_M, N) \longrightarrow \hom_\Lambda(P_{M'}', N) \longrightarrow 0 .$$

Then apply the dual of theorem 2.3.   □

We can also define the extension functor as follows: Let I_N be a reduced projective resolution of N. Consider $\hom_\Lambda(M, I_N)$ which is a semiexact sequence whose n-th cohomology $H^n(\hom_\Lambda(M, I_N))$ will be denoted by $\overline{\operatorname{Ext}}_\Lambda^n(M, N)$.
    Some important results for $\operatorname{Ext}_\Lambda^n(\_, \_)$ and $\overline{\operatorname{Ext}}_\Lambda^n(\_, \_)$ are the following:

Theorem 3.11. *a) The functors* $\operatorname{Ext}_\Lambda^0(\_, N)$ *and* $\overline{\operatorname{Ext}}_\Lambda^0(M, \_)$ *are naturally equivalent to the functors* $\hom_\Lambda(\_, N)$ *and* $\hom_\Lambda(M, \_)$ *respectively.*
b) Let I *be an injective* Λ-*module. Then* $\operatorname{Ext}_\Lambda^n(M, I) = 0$ *for all* $n \in \mathbb{Z}^+$ *and any* M.
c) Let P *be a projective* Λ-*module. Then* $\operatorname{Ext}_\Lambda^n(P, N) = 0$ *for all* $n \in \mathbb{Z}^+$ *and any* N.

A corresponding theorem to 3.6 is

Theorem 3.12. $\operatorname{Ext}_\Lambda^n(M, N) \cong \overline{\operatorname{Ext}}_\Lambda^n(M, N)$.

Recall that an element x of a Λ-module N for Λ an integral domain is a *torsion element* if there is an $\alpha \in \Lambda$ different from 0 such that $\alpha x = 0$. The set of torsion elements form a submodule τ_N of N called the *torsion submodule* of N. A Λ-module N is called a *torsion module* if $\tau_N = N$ and is called *torsion free* if $\tau_N = 0$.
    $\tau_\_$ can be viewed as a functor from the category of Λ-modules (Λ an integral domain) to the category of torsion modules.
    Let Λ be an integral domain and F its quotient field. Consider the exact sequence

$$0 \longrightarrow \Lambda \longrightarrow F \longrightarrow F/\Lambda \longrightarrow 0 .$$

A justification for the name Tor is the following result:

Theorem 3.13. *a) The functors τ_- and $\mathrm{Tor}_1^\Lambda(F/\Lambda, \_)$ are naturally equivalent for Λ an integral domain.*
b) If Λ is an integral domain then $\mathrm{Tor}_n^\Lambda(M, N)$ is a torsion Λ-module for all $n \geq 1$.

Consider the following problem: let M' and M'' be two Λ-modules. Which are the Λ-modules M such that M' be a submodule of M and M'' be its quotient? Equivalently, which are the Λ-modules M such that the sequence

$$0 \longrightarrow M' \longrightarrow M \longrightarrow M'' \longrightarrow 0$$

is exact? The classification of such Λ-modules M constitute what is known as the *extension problem.*

An *extension* of M' by M'' is a short exact sequence

$$E : M' \rightarrowtail M \twoheadrightarrow M'' \ .$$

We will say that two extensions of M' by M'', $M' \rightarrowtail M_1 \twoheadrightarrow M''$ and $M' \rightarrowtail M_2 \twoheadrightarrow M''$, are equivalent if there is a homomorphism $\psi : M_1 \longrightarrow M_2$ such that the following diagram commutes

$$
\begin{array}{ccccc}
M' & \rightarrowtail & M_1 & \twoheadrightarrow & M' \\
1_{M'} \downarrow & & \downarrow \psi & & \downarrow 1_{M''} \\
M' & \rightarrowtail & M_2 & \twoheadrightarrow & M''
\end{array}
$$

ψ must be an isomorphism. Denote by $\mathrm{Ex}(M'', M')$ the set of equivalence classes of extensions of M' by M''. Let $E_1 : M' \rightarrowtail M_1 \twoheadrightarrow M''$ and $E_2 : M' \rightarrowtail M_2 \twoheadrightarrow M''$ be two extensions of M' and M''. Define its sum as the extension

$$E_1 \oplus E_2 : M' \oplus M' \rightarrowtail M_1 \oplus M_2 \twoheadrightarrow M'' \oplus M'' \ .$$

Let $\nabla_{M''} : M'' \oplus M'' \longrightarrow M''$ and $\triangle_{M'} : M' \longrightarrow M' \oplus M'$ be homomorphisms given by $\nabla_{M''}(x_1, x_2) = x_1 + x_2$ and $\triangle_{M'}(y) = (y, y)$. We define the sum of two elements in $\mathrm{Ex}(M'', M')$ as

$$E_1 + E_2 = \nabla_{M''}(E_1 \oplus E_2)\triangle_{M'} \ ,$$

which we call the *Baer sum.* With this Baer sum it can be proved that $\mathrm{Ex}(M'', M')$ adquires an abelian group structure whose neutral element is the class of the split extension.

The following theorem justifies the name Ext:

Theorem 3.14. *The groups $\mathrm{Ex}(M'', M')$ and $\mathrm{Ext}_\Lambda^1(M'', M')$ are isomorphic.*

Example 3.15. Let $\Lambda = \mathbb{Z}$. Since $\mathbb{Z}$ is projective, for $n \geq 1$

$$\mathrm{Ext}_{\mathbb{Z}}^n(\mathbb{Z}, \mathbb{Z}) = 0 \quad \text{and} \quad \mathrm{Ext}_{\mathbb{Z}}^n(\mathbb{Z}, \mathbb{Z}/n) = 0 \ .$$

This means that the extensions of $\mathbb{Z}$ by $\mathbb{Z}$ or of $\mathbb{Z}/n$ by $\mathbb{Z}$ split and contain only one element.

Exercise 3.16. Verify that $\text{Ext}^1_{\mathbb{Z}}(\mathbb{Z}/n, \mathbb{Z}) = \mathbb{Z}/n$ and $\text{Ext}^1_{\mathbb{Z}}(\mathbb{Z}/n, \mathbb{Z}/m) = \mathbb{Z}/(n, m)\mathbb{Z}$.

II. Cohomology of Groups

II.1. G-Modules and the (Co)Homology of a Group

Let G be a group. We will associate to G a ring denoted by $\mathbb{Z}[G]$.

Definition 1.1. The integral group ring $\mathbb{Z}[G]$ of the multiplicative group G is the set of formal sums $\sum_{i \in I} \lambda_i g_i$, $\lambda_i \in \mathbb{Z}$, $g_i \in G$ where almost any λ_i is zero, together with the two binary operations $+, \cdot$ of $\mathbb{Z}[G]$ defined as follows:

$$\left(\sum_{i \in I} \lambda_i g_i \right) + \left(\sum_{i \in I} \mu_i g_i \right) = \sum_{i \in I} (\lambda_i + \mu_i) g_i$$

$$\left(\sum_{i \in I} \lambda_i g_i \right) \cdot \left(\sum_{i \in I} \mu_i g_i \right) = \sum_{i \in I} \left(\sum_{g_j g_k = g_i} \lambda_j \mu_k \right) g_i \;.$$

It is immediate to check that $(\mathbb{Z}[G], +, \cdot)$ is a ring. Also $(\mathbb{Z}[G], \cdot)$ contains G as a subgroup. Then if G is not abelian, $\mathbb{Z}[G]$ is not a commutative ring. Briefly, $\mathbb{Z}[G]$ consists of the free abelian group generated by the elements of G as basis such that the product of two elements is given by the product of G.

The ring $\mathbb{Z}[G]$ has the following universal property:

Theorem 1.2. *Let Λ be a ring with 1_{Λ} and $\varphi : G \longrightarrow \Lambda$ a function such that $\varphi(1) = 1_{\Lambda}$ and $\varphi(g_i g_j) = \varphi(g_i)\varphi(g_j)$. Then there exists a unique ring homomorphism $\psi : \mathbb{Z}[G] \longrightarrow \Lambda$ such that the following diagram commutes*

$$
\begin{array}{ccc}
G & \xrightarrow{\;\iota\;} & \mathbb{Z}[G] \\
 & {\scriptstyle \varphi} \searrow & \downarrow {\scriptstyle \psi} \\
 & & \Lambda
\end{array}
$$

Proof. Define $\psi \left(\sum_{i \in I} \lambda_i g_i \right) = \sum_{i \in I} \lambda_i \varphi(g_i)$. Clearly ψ is the unique ring homomorphism such that $\varphi = \psi \circ \iota$. $\qquad \square$

Consider the case where φ is the trivial function $\varphi : G \longrightarrow \mathbb{Z}$ which sends every element $g \in G$ into $1 \in \mathbb{Z}$. By 1.2, φ gives rise to a unique ring homomorphism $\varepsilon : \mathbb{Z}[G] \longrightarrow \mathbb{Z}$ which we may call the *augmentation* of $\mathbb{Z}[G]$. Then if $\sum_{i \in I} \lambda_i g_i \in \mathbb{Z}[G]$, $\varepsilon \left(\sum_{i \in I} \lambda_i g_i \right) = \sum_{i \in I} \lambda_i$, $(\varepsilon(g_i) = 1)$.
We will call the kernel of ε the *augmentation ideal* of G and denote it by IG.

Definition 1.3. Let $(M, +)$ be an abelian group. M is a left G-*module* if there is an action of G on M such that

  i) $1x = x$; $x \in M$.

  ii) $(gg')x = g(g'x)$, $g, g' \in G$; $x \in M$.

  iii) $g(x_1 + x_2) = gx_1 + gx_2$; $g \in G$; $x_1, x_2 \in M$.

In other words, a G-module M consists of an abelian group M together with a homomorphism $\kappa : G \longrightarrow \operatorname{Aut}(M)$. By theorem 1.2 $\kappa : G \longrightarrow \Lambda = \operatorname{Aut}(M) \subset \operatorname{End}(M)$ determines a unique homomorphism $\varphi : \mathbb{Z}[G] \longrightarrow \operatorname{End}(M)$ giving M a left $\mathbb{Z}[G]$-module structure. Also, since any ring homomorphism sends invertible elements into invertible elements and since the elements of the additive part of $\mathbb{Z}[G]$ are invertible, if M is a left $\mathbb{Z}[G]$-module then it is a left G-module. Then we will speak indistinctively of a left $\mathbb{Z}G$-module M or a left G-module M.

A G-module is called *trivial* if $gx = x$ for all $g \in G$, $x \in M$, i.e. all the elements of M remain fixed under the action of G.

Now we will consider a special case of I.3.1 where $\Lambda = \mathbb{Z}[G]$, $P_{\mathbb{Z}}$ a reduced G-projective resolution of the trivial G-module $\mathbb{Z}$ and N a left G-module.

Definition 1.4. The n-th homology group of a group G with coefficients in a left G-module N is
$$H_n(G; N) = \operatorname{Tor}_n^{\mathbb{Z}G}(\mathbb{Z}, N) \ .$$

Also we can consider a special case of I.3.7 where $\Lambda = \mathbb{Z}[G]$, $P_{\mathbb{Z}}$ a reduced G-projective resolution of the trivial G-module $\mathbb{Z}$ and N a left G-module.

Definition 1.5. The n-th cohomology group of a group G with coefficients in a left G-module N is
$$H^n(G; N) = \operatorname{Ext}_{\mathbb{Z}G}^n(\mathbb{Z}, N) \ .$$

Since $\operatorname{Tor}_n^{\mathbb{Z}G}(\mathbb{Z}, \_)$ and $\operatorname{Ext}_{\mathbb{Z}G}^n(\mathbb{Z}, \_)$ are covariant functors, $H_n(G, \_)$ and $H^n(G, \_)$ are covariant functors from the category of G-modules to the category of abelian groups. Also we will have long exact sequences in the second variable only.

Theorem 1.6. *Let* $N' \rightarrowtail N \twoheadrightarrow N''$ *be a short exact sequence of G-modules. Then there are homomorphisms*

$$\kappa_n : H_n(G; N'') \longrightarrow H_{n-1}(G; N')$$

and

$$\kappa^n : H^n(G; N'') \longrightarrow H^{n+1}(G; N')$$

for every $n \in \mathbb{Z}$, *such that the following sequences are exact*

$$\ldots \longrightarrow H_n(G; N') \longrightarrow H_n(G; N) \longrightarrow H_n(G; N'') \xrightarrow{\kappa_n} H_{n-1}(G; N') \longrightarrow \ldots$$

$$\ldots \longrightarrow H^n(G; N') \longrightarrow H^n(G; N) \longrightarrow H^n(G; N'') \xrightarrow{\kappa^n} H^{n+1}(G; N') \longrightarrow \ldots \ .$$

Proof. Apply I.3.3 and I.3.9.   □

Proposition 1.7. *Let* P *be a projective* $\mathbb{Z}G$-*module and* I *an injective* $\mathbb{Z}G$-*module. Then*

$$H_n(G; P) = 0 \quad \text{for} \quad n \geq 1$$

and

$$H^n(G; I) = 0 \quad \text{for} \quad n \geq 1 \ .$$

Proof. Apply theorems I.3.5(b) and I.3.11(b).   □

Definition 1.8. Let G be a group and N a G-module. The *group of invariants of* N, denoted by N_G, is the quotient of N by the additive subgroup generated by the elements of the form $gy - y$, $g \in G$, $y \in N$, i.e. $N_G = N/T$ where $T = < gy - y >$. The *subgroup of invariants of* N, denoted by N^G, consists of all the elements $y \in N$ such that the action of G is trivial; i.e. $N^G = \{y \in N \mid gy = y \ \text{for all} \ g \in G\}$.

Observe that N^G is the largest submodule in which G acts trivially and N_G is largest of the quotients of N in which the action of G over N is trivial.

Since $gy - y = (g - 1)y$ and the elements $(g - 1) \in \mathbb{Z}G$ generate IG we will write $T = IG \circ N$. If we apply the functor $\_ \otimes_{\mathbb{Z}G} N$ to the short exact sequence $IG \rightarrowtail \mathbb{Z}G \twoheadrightarrow \mathbb{Z}$ we obtain the exact sequence

$$IG \otimes_{\mathbb{Z}G} N \longrightarrow \mathbb{Z}G \otimes_{\mathbb{Z}G} N \longrightarrow \mathbb{Z} \otimes_{\mathbb{Z}G} N \ .$$

Then $\mathbb{Z}G \otimes_{\mathbb{Z}G} N \cong N$ and under this isomorphism, im $(IG \otimes_{\mathbb{Z}G} N)$ goes to $IG \circ N$. Thus $\mathbb{Z} \otimes_{\mathbb{Z}G} N \cong N/IG \circ N = N/T = N_G$.

Theorem 1.9. *Let* G *be a group and* N *a* G-*module. Then*

$$H_0(G; N) = N_G \quad \text{and} \quad H^0(G; N) = N^G \ .$$

Proof. By definition, $H_0(G; N) = \text{Tor}_0^{\mathbb{Z}G}(\mathbb{Z}, N) = \mathbb{Z} \otimes_{\mathbb{Z}G} N$. Hence $H_0(G; N) = N_G$. Now, $H^0(G; N) = \text{Ext}_{\mathbb{Z}G}^0(\mathbb{Z}, N) = \text{Hom}_{\mathbb{Z}G}(\mathbb{Z}, N)$ by definition. But a homomorphism $\varphi : \mathbb{Z} \longrightarrow N$ is completely determined by $\varphi(1) = y \in N$. Since φ is a G-module homomorphism, $g \circ y = \varphi(g \circ 1) = \varphi(1) = y$ for all $g \in G$, i.e. φ

is a G-module homomorphism if and only if $\varphi(1) = y$ remains fixed under the G-action. Then $H^0(G; N) = N^G$.   □

Corollary 1.10. *If N is a trivial G-module then $H_0(G; N) = N$ and $H^0(G; N) = N$.*

II.2. $H_1(G, N)$ and $H^1(G, N)$

In this section we will study the (co)homology of degree one of a group G with coefficients in a G-module N.

Lemma 2.1. *Let G be a group, IG its augmentation ideal. Then the additive group $IG/(IG)^2$ is isomorphic to the multiplicative group $G/[G, G]$ where $[G, G]$ is the commutator subgroup of G.*

Proof. The reader can prove that the elements of the form $g - 1$, $1 \neq g \in G$ form a basis of IG.

Define
$$\varphi : G \longrightarrow IG/(IG)^2$$
by $\varphi(g) = (g - 1) + (IG)^2$. It is easy to see that φ is a homomorphism and since $[G, G] \subset \ker \varphi$ and $IG/(IG)^2$ is abelian, φ induces $\psi : G/[G, G] \longrightarrow IG/(IG)^2$ where
$$g[G, G] \longmapsto (g - 1) + (IG)^2 \ .$$

Now, define $\varphi' : IG \longrightarrow G/[G, G]$ by $\varphi'(g - 1) = g[G, G]$. If $x \in (IG)^2$ it is easy to see that $x \in \ker \varphi'$. Hence φ' induces $\psi' : IG/(IG)^2 \longrightarrow G/[G, G]$. Finally ψ and ψ' are mutually inverse.   □

Theorem 2.2. $H_1(G, \mathbb{Z}) \cong IG/(IG)^2$.

Proof. By definition 1.4, $H_1(G, \mathbb{Z}) = \text{Tor}_1^{\mathbb{Z}G}(\mathbb{Z}, \mathbb{Z})$. Consider the projective presentation of $\mathbb{Z}$, $IG \rightarrowtail \mathbb{Z}G \twoheadrightarrow \mathbb{Z}$. Apply the functor $\mathbb{Z} \otimes_{\mathbb{Z}G} \_$ to it and obtain by the results of §I.3 the following exact sequence

$$\ldots \longrightarrow \text{Tor}_1^{\mathbb{Z}G}(\mathbb{Z}, IG) \longrightarrow \text{Tor}_1^{\mathbb{Z}G}(\mathbb{Z}, \mathbb{Z}G) \longrightarrow \text{Tor}_1^{\mathbb{Z}G}(\mathbb{Z}, \mathbb{Z}) \xrightarrow{\kappa_1}$$
$$\xrightarrow{\kappa_1} \mathbb{Z} \otimes_{\mathbb{Z}G} IG \longrightarrow \mathbb{Z} \otimes_{\mathbb{Z}G} \mathbb{Z}G \xrightarrow{\varphi_0} \mathbb{Z} \otimes_{\mathbb{Z}G} \mathbb{Z} \longrightarrow 0 \ .$$

Since $\mathbb{Z}G$ is projective, $H_1(G, \mathbb{Z}G) = \text{Tor}_1^{\mathbb{Z}G}(\mathbb{Z}, \mathbb{Z}G) = 0$. Since $\mathbb{Z}$ is a trivial G-module, $H_0(G, \mathbb{Z}G) = \text{Tor}_0^{\mathbb{Z}G}(\mathbb{Z}, \mathbb{Z}G) = \mathbb{Z} \otimes_{\mathbb{Z}G} \mathbb{Z}G \cong \mathbb{Z}$ and $H_0(G, \mathbb{Z}) = \text{Tor}_0^{\mathbb{Z}G}(\mathbb{Z}, \mathbb{Z}) = \mathbb{Z} \otimes_{\mathbb{Z}G} \mathbb{Z} \cong \mathbb{Z}$. Hence the homomorphism

$$\varphi_{0_*} : H_0(G, \mathbb{Z}G) \longrightarrow H_0(G, \mathbb{Z})$$

is onto and $\varphi_{0_*} \neq 0$. Any endomorphism $\varphi_0 : \mathbb{Z} \longrightarrow \mathbb{Z}$ is either monomorphic or trivial. Since the induced φ_{0_*} is different from 0, φ_0 is monomorphic. By exactness, $\kappa_1 : \mathrm{Tor}_1^{\mathbb{Z}G}(\mathbb{Z}, \mathbb{Z}) \longrightarrow \mathrm{Tor}_0^{\mathbb{Z}G}(\mathbb{Z}, IG) = \mathbb{Z} \otimes_{\mathbb{Z}G} IG$ is an isomorphism. By 1.4 and 1.9,

$$\kappa_1 : H_1(G, \mathbb{Z}) \xrightarrow{\simeq} H_0(G, IG) = (IG)_G = IG/IG \circ IG .$$

$\square$

Corollary 2.3. $H_1(G, \mathbb{Z}) \cong G/[G, G]$.

Now we will consider $H^1(G, \mathbb{Z})$. By definition

$$H^1(G, \mathbb{Z}) = \mathrm{Ext}_{\mathbb{Z}G}^1(\mathbb{Z}, \mathbb{Z}) .$$

Applying the contravariant functor $\hom_{\mathbb{Z}G}(\_, \mathbb{Z})$ to the projective presentation of $\mathbb{Z}$, $IG \rightarrowtail \mathbb{Z}G \twoheadrightarrow \mathbb{Z}$ and using the results of §I.3 we obtain that $H^1(G, \mathbb{Z}) \cong \hom_{\mathbb{Z}G}(IG, \mathbb{Z})$. But a homomorphism $f : IG \longrightarrow \mathbb{Z}$ is a G-module homomorphism if and only if $f(x(y-1)) = xf(y-1) = f(y-1)$; $x, y \in G$, i.e. if any only if $f((x-1)(y-1)) = 0$. Hence, using 2.1 we obtain the following

Theorem 2.4. $H^1(G, \mathbb{Z}) \cong \hom_{\mathbb{Z}}(IG/(IG)^2, \mathbb{Z}) \cong \hom_{\mathbb{Z}}(G/[G, G], \mathbb{Z})$.

Observe that since $G/[G, G] \cong H_1(G, \mathbb{Z})$,

$$H^1(G, \mathbb{Z}) \cong \hom_{\mathbb{Z}}(H_1(G, \mathbb{Z}), \mathbb{Z}) .$$

Let N be a trivial G-module. Then the reader might prove that

$$H_1(G, N) \cong N \otimes_{\mathbb{Z}} G/[G, G] \qquad \text{and}$$
$$H^1(G, N) \cong \hom_{\mathbb{Z}}(G/[G, G], N) .$$

We will devote the rest of this section to interpret $H^1(G, N)$ when N is not a trivial G-module. For this we will introduce the concept of derivation and establish a natural equivalence between the functors $\hom_{\mathbb{Z}G}(IG, \_)$ and $\mathrm{Der}(G, \_)$ which will allow us to obtain such interpretation.

Definition 2.5. A *derivation or crossed homomorphism* is a function $f : G \longrightarrow N$ where $(G, \cdot)$ is a group and N is a G-module such that

$$f(x \cdot y) = x \circ f(y) + f(x)$$

where $\circ$ denotes the action of G on N.

If N is a trivial G-module, then f is a homomorphism from the multiplicative group G into the abelian group N. Observe that $f(1) = 0$. If we define the sum of two derivations f and g as $(f + g)(x) = f(x) + g(x)$ it is clear that $f + g$ is a derivation. We will denote by $\mathrm{Der}(G, N)$ the set of all derivations from G to N together with the sum of derivations. It is easy to check that $\mathrm{Der}(G, N)$ is an abelian group under such sum.

Consider the set of derivations $f : G \longrightarrow N$ of the form $f_a(x) = xa - a$ with $a \in N$ fixed. Such set is called the set of *principal derivations* and is denoted by $\mathrm{PDer}(G, N)$. Since $f_a + f_{a'} = f_{(a+a')}$ and $f_{(-a)} = -f_{a'}$, $\mathrm{PDer}(G, N)$ is a subgroup of $\mathrm{Der}(G, N)$.

We leave to the reader to prove the following

Proposition 2.6. $\mathrm{Der}(G, \_) : \mathrm{Mod}_{\mathbf{Z}G} \longrightarrow Ab$ *is a covariant functor which is naturally equivalent to* $\mathrm{Hom}_{\mathbf{Z}G}(IG, \_)$.

Theorem 2.7. *Let G be a group and N a G-module. Then*

$$H^1(G, N) \cong \mathrm{Der}(G, N)/\mathrm{PDer}(G, N) \ .$$

Proof. By definition, $H^1(G, N) = \mathrm{Ext}^1_{\mathbf{Z}G}(\mathbf{Z}, N)$. Consider the $\mathbf{Z}G$-free presentation of $\mathbf{Z}$, $IG \overset{i}{\rightarrowtail} \mathbf{Z}G \overset{p}{\twoheadrightarrow} \mathbf{Z}$. By the results of I.3 we obtain a long exact sequence

$$0 \longrightarrow \hom_{\mathbf{Z}G}(\mathbf{Z}, N) \overset{p^*}{\longrightarrow} \hom_{\mathbf{Z}G}(\mathbf{Z}G, N) \overset{i^*}{\longrightarrow} \hom_{\mathbf{Z}G}(IG, N) \longrightarrow$$
$$\longrightarrow \mathrm{Ext}^1_{\mathbf{Z}G}(\mathbf{Z}, N) \longrightarrow \mathrm{Ext}^1_{\mathbf{Z}G}(\mathbf{Z}G, N) \longrightarrow \ldots \ .$$

Since $\hom_{\mathbf{Z}G}(\mathbf{Z}G, N) \cong N$ and $\mathrm{Ext}^1_{\mathbf{Z}G}(\mathbf{Z}G, N) = 0$ since $\mathbf{Z}G$ is free, we have that

$$0 \longrightarrow \hom_{\mathbf{Z}G}(\mathbf{Z}, N) \overset{p^*}{\longrightarrow} N \overset{i^*}{\longrightarrow} \hom_{\mathbf{Z}G}(IG, N) \longrightarrow H^1(G, N) \longrightarrow 0$$

is exact. Then $H^1(G, N) \cong \mathrm{coker}\, i^* = \hom_{\mathbf{Z}G}(IG, N)/i^*N$ where $i^*(a)(x-1) = xa - a$, $a \in N$, $x \in G$. By 2.6 there is a derivation $f_{i^*} : G \longrightarrow N$ associated to $i^*(a)$ of the form $f_{i^*}(x) = (x-1)a$. Such derivations are the principal derivations $\mathrm{PDer}(G, N)$. By 2.6 we have

$$H^1(G, N) = \hom_{\mathbf{Z}G}(IG, N)/i^*N \cong \mathrm{Der}(G, N)/\mathrm{PDer}(G, N) \ .$$

□

Observe that what 3.7 tells us is that $H^1(G, N)$ measures the size of the derivations that are not principal.

Exercise 2.8. Prove that if F is a free group on the free generating set S, then IF is a free $\mathbf{Z}F$-module on the basis $S - 1$.

II.3. $H_2(G, N)$ and Some Calculations

We will mention a very important lemma which gives a *five term exact sequence* for homology:

Lemma 3.1. *Let* $H \rightarrowtail G \twoheadrightarrow Q$ *be a short exact sequence of groups and* N *a left* Q-module. *Then the following sequence is exact:*

$$H_2(G, N) \longrightarrow H_2(Q, N) \longrightarrow N \otimes_{\mathbb{Z}Q} H/[H, H]$$
$$\longrightarrow N \otimes_{\mathbb{Z}G} IG \longrightarrow N \otimes_{\mathbb{Z}Q} IQ \longrightarrow 0 \ .$$

Consider a free presentation of a group G, $R \rightarrowtail F \twoheadrightarrow G$. By 3.1 there is a five term exact sequence

$$H_2(F, N) \longrightarrow H_2(G, N) \longrightarrow N \otimes_{\mathbb{Z}G} R/[R, R]$$
$$\longrightarrow N \otimes_{\mathbb{Z}F} IF \longrightarrow N \otimes_{\mathbb{Z}G} IG \longrightarrow 0 \ . \tag{3.2}$$

It is easy to prove that (see Exercise 3.10), $H_2(F, N) = 0$, (F a free group) hence $H_2(G, N) \cong \ker (N \otimes_{\mathbb{Z}G} R/[R, R] \longrightarrow N \otimes_{\mathbb{Z}G} IF)$.

It happens that for $N = \mathbb{Z}$, the exact sequence of 3.1 transforms into

$$H_2(G, \mathbb{Z}) \longrightarrow H_2(Q, \mathbb{Z}) \longrightarrow H/[G, H] \longrightarrow G/[G, G] \longrightarrow Q/[Q, Q] \longrightarrow 0$$

where $[G, H]$ denotes the normal subgroup of H generated by elements of the form $ghg^{-1}h^{-1}$, $g \in G$, $h \in H$.

Returning to 3.2, for $N = \mathbb{Z}$ we obtain

$$\begin{aligned}
H_2(G, \mathbb{Z}) &\cong \ker (\mathbb{Z} \otimes_{\mathbb{Z}G} R/[R, R] \longrightarrow \mathbb{Z} \otimes_{\mathbb{Z}F} IF) \\
&\cong \ker(R/[F, R] \longrightarrow F/[F, F]) \\
&\cong (R \cap [F, F])/[F, R] \ .
\end{aligned}$$

Now, we will see another situation where $H_2(G, \mathbb{Z})$ appears.

Definition 3.3. A *central extension of* G is a short exact sequence of groups

$$1 \longrightarrow K \longrightarrow E \longrightarrow G \longrightarrow 1$$

such that $K \subset Z(E)$ where $Z(E)$ denotes the center of E.

Definition 3.4. A *universal central extension* of G is a central extension

$$1 \longrightarrow N \longrightarrow U \longrightarrow G \longrightarrow 1$$

such that, given any central extension

$$1 \longrightarrow K \longrightarrow E \longrightarrow G \longrightarrow 1$$

there is a unique homomorphism $h : U \longrightarrow E$ such that the following diagram commutes:

$$
\begin{array}{ccccccccc}
1 & \longrightarrow & N & \longrightarrow & U & \longrightarrow & G & \longrightarrow & 1 \\
 & & \downarrow & & \downarrow h & & \| & & \\
1 & \longrightarrow & K & \longrightarrow & E & \longrightarrow & G & \longrightarrow & 1
\end{array}
$$

A group G is called *perfect* if $G = [G,G]$. Let G be a perfect group and choose a homomorphism $F \twoheadrightarrow G$ where F is a free group. Let $R = \ker(F \twoheadrightarrow G)$. Then $[R, F]$ is a normal subgroup of F. Since $G \cong F/R$, there is an epimorphism $\phi : F/[R, F] \longrightarrow G$ such that $\ker \phi \subset Z(F/[R, F])$. Then $[F/[R, F], F/[R, F]] \cong [F, F]/[R, F]$ is a perfect central extension of G. Let $1 \longrightarrow N \longrightarrow U \longrightarrow G \longrightarrow 1$ be any other central extension of G. Since F is free, there is a homomorphism $h : F \longrightarrow U$ such that the following diagram commutes:

$$
\begin{array}{ccc}
F & \longrightarrow & G \\
h \downarrow & & \| \\
U & \longrightarrow & G
\end{array}
$$

Since $1 \longrightarrow N \longrightarrow U \longrightarrow G \longrightarrow 1$ is a central extension $h([R, F]) = 1$. Then h induces a homomorphism

$$F/[R, F] \longrightarrow U$$

and if we restrict it to $[F, F]/[R, F]$ we will have a homomorphism $[F, F]/[R, F] \longrightarrow U$ which is unique. Hence

$$1 \longrightarrow \ker \phi \longrightarrow [F, F]/[R, F] \overset{\phi}{\longrightarrow} G \longrightarrow 1$$

is a universal central extension of G. But $\ker \phi$ is none other that $(R \cap [F, F])/[R, F]$. And this means that $\ker \phi = H_2(G, \mathbb{Z})$. Finally, we can say that $H_2(G, \mathbb{Z})$ is the kernel of the universal central extension

$$1 \longrightarrow H_2(G, \mathbb{Z}) \longrightarrow [F, F]/[R, F] \longrightarrow G \longrightarrow 1 \ .$$

Here are some calculations of homology and cohomology of diverse groups.

Example 3.5. Let $G = \mathbb{Z}/n$ the integers modulo n. Then by 2.3, $H_1(\mathbb{Z}/n, \mathbb{Z}) \cong \mathbb{Z}/n$ since $\mathbb{Z}/n$ is abelian and hence $[\mathbb{Z}/n, \mathbb{Z}/n] = \{1\}$.

Example 3.6. Let $G = \mathbb{Z}/n$. Then by 2.4,

$$H^1(\mathbb{Z}/n, \mathbb{Z}) \cong \hom_{\mathbb{Z}} (H_1(\mathbb{Z}/m, \mathbb{Z}), \mathbb{Z}) = \hom_{\mathbb{Z}}(\mathbb{Z}/n, \mathbb{Z}) = 0 \ .$$

Exercise 3.7. Let C_n denote a cyclic group of order n with generator g. Prove that

$$W : \ldots \overset{D}{\longrightarrow} \mathbb{Z}C_n \overset{N}{\longrightarrow} \mathbb{Z}C_n \overset{D}{\longrightarrow} \mathbb{Z}C_n \overset{\epsilon}{\longrightarrow} \mathbb{Z} \longrightarrow 0$$

where $D = g - e$ and $N = e + g + g^2 + \ldots + g^{n-1}$ is a free C_n-resolution of $\mathbb{Z}$.

Example 3.8. Let M be a C_n-module. If you apply the functors $\_ \otimes_{\mathbb{Z}C_n} M$ and $\hom_{\mathbb{Z}C_n}(\_, M)$ to the resolution W of 3.7 and then calculate its homology and cohomology respectively you will obtain

$$H_i(C_n, M) = \begin{cases} M_{C_n} & i = 0 \\ \ker D/N(M) & i = 2n - 1 \\ \ker N/D(M) & i = 2n \end{cases}$$

and

$$H^i(C_n, M) = \begin{cases} M^{C_n} & i = 0 \\ \ker N/D(M) & i = 2n - 1 \\ \ker D/N(M) & i = 2n \ . \end{cases}$$

Example 3.9. Let C denote an infinitive cyclic group with generator g. Consider the resolution of $\mathbb{Z}$ over $\mathbb{Z}C$

$$0 \longrightarrow \mathbb{Z} \overset{g-e}{\longrightarrow} \mathbb{Z} \overset{\varepsilon}{\longrightarrow} \mathbb{Z} \longrightarrow 0 \ .$$

Let M be a C-module and apply the functors $\_ \otimes_{\mathbb{Z}C} M$ and $\hom_{\mathbb{Z}C}(\_, M)$ to this resolution and calculate its homology and cohomology respectively. Then you will obtain

$$H_i(C, M) = \begin{cases} M_C & i = 0 \\ M^C & i = 1 \\ 0 & i \neq 0, 1 \end{cases}$$

and

$$H^i(C, M) = \begin{cases} M^C & i = 0 \\ M_C & i = 1 \\ 0 & i \neq 0, 1 \ . \end{cases}$$

Exercise 3.10. Prove that if F is a free group

$$H_n(F, N) = H^n(F, M) = 0 \quad \text{for} \quad n \geq 2 \ .$$

(Use Exercise 2.8 in order to take $IF \rightarrowtail \mathbb{Z}F \twoheadrightarrow \mathbb{Z}$ as a $\mathbb{Z}F$-free resolution of $\mathbb{Z}$.)

See Snaith's lectures, Chapter II, for other specific examples.

III. Classical Algebraic K-Theory

III.1. $K_0\Lambda$

Let Λ be any ring, not necessarily commutative and $_\Lambda$Mod the category of left Λ-modules. Let $\mathcal{C}$ be a subcategory of $_\Lambda$Mod. The *Grothendieck construction* consists in associating to certain subcategory of $_\Lambda$Mod a certain abelian group which we will denote by $K_0\mathcal{C}$.

Specifically, let M be a Λ-module $\mathcal{C}$. Let $\langle M \rangle$ denote the isomorphism class of M. Let F be the abelian group with basis $\{\langle M \rangle \mid M \in \mathcal{C}\}$ and let R denote the subgroup of F generated by the expressions of the form $\langle M \rangle - \langle M' \rangle - \langle M'' \rangle$ where $M' \rightarrowtail M \twoheadrightarrow M''$ runs over all short exact sequences in $\mathcal{C}$. Then we define

$$K_0\mathcal{C} = F/R$$

and call it *the Grothendieck group of $\mathcal{C}$*.

We will denote by $[M]$ the image of $\langle M \rangle$ in $K_0\mathcal{C}$. Hence every time we have a short exact sequence in $\mathcal{C}$, $M' \rightarrowtail M \twoheadrightarrow M''$, we will have an expression of the form $[M] = [M'] + [M'']$ in $K_0\mathcal{C}$. This means that $K_0\mathcal{C}$ is generated by $\{[M] \mid M \in \mathcal{C}\}$ subject to the relations of the form $[M] = [M'] + [M'']$.

Let's apply the Grothendieck construction to a specific subcategory of $_\Lambda$Mod: let $_\Lambda P$ be the category of left finitely generated projective Λ-modules. By abuse in the notation we will write $K_0(_\Lambda P)$ simply as $K_0\Lambda$. Then we have the following

Definition 1.1. $K_0\Lambda$ is the abelian group F/R where F is a free abelian group whose generators $\langle P \rangle$ are the isomorphism classes of finitely generated projective Λ-modules and R is the subgroup of F generated by the expressions of the form

$$\langle P \rangle + \langle Q \rangle - \langle P \oplus Q \rangle \ .$$

$K_0\Lambda$ is called the *projective class group* of Λ.

It is easy to prove the following universal property satisfied by K_0:

Proposition 1.2. *Let $\phi : \mathcal{C} \longrightarrow G$ be a function from $\mathcal{C}$ to an additive abelian group G such that for $M \in \mathcal{C}$, $\phi(\langle M \rangle)$ depends only on the isomorphism class of M and $\phi(\langle M \rangle) = \phi(\langle M' \rangle) + \phi(\langle M'' \rangle)$. Then there is a unique group homomorphism $\psi : K_0\mathcal{C} \longrightarrow G$ such that $\phi(\langle M \rangle) = \psi([M])$.*

Example 1.3. Let k be a field and $\mathcal{V}$ the subcategory of vector spaces of finite dimension over k of the category $_k$Mod, i.e. of the category of finitely generated projective k-modules. The function $\phi : \mathcal{V} \longrightarrow \mathbb{Z}$ given by $\langle V \rangle \longmapsto \dim V$ induces $\psi : K_0\mathcal{V} = K_0 k \longrightarrow \mathbb{Z}$ since $R \subset \ker \phi$ because $\dim(V \oplus W) = \dim V + \dim W$. ψ is given by the rule $[V] \longmapsto \dim V$. Since $\psi([k]) = 1$, ψ is onto. If $\psi([V] - [W]) = 0$ then $\dim V = \dim W$ and $V \cong W$, hence ψ is injective. So $K_0 k = K_0 \mathcal{V} \cong \mathbb{Z}$.

Observe that K_0 of a field measures, to a certain extent, how much do the finitely generated projective Λ-modules lack a dimension theory like the vector spaces. In this way we can see the part concerning K_0 of algebraic K-theory as an effort to generalize certain elementary properties of linear algebra to modules over any ring.

In case the ring Λ is commutative, the tensor product over Λ of two Λ-modules is a Λ-module by means of $\lambda(x \otimes_\Lambda y) = \lambda x \otimes_\Lambda y = x \otimes_\Lambda \lambda y$. Since $\Lambda^n \otimes \Lambda^m \cong \Lambda^{nm}$, the category of finitely generated projective Λ-modules $_\Lambda P$ is closed under $\otimes$. Then we can give $K_0\Lambda$ a ring structure using $[P] \cdot [Q] = [P \otimes_\Lambda Q]$. This means, formally, that $K_0\Lambda = F/R$ can have a ring structure defining $\langle P \rangle \langle Q \rangle = \langle P \otimes_\Lambda Q \rangle$. R turns into an ideal of F, so $K_0\Lambda$ is a ring.

Let $F : \Lambda \longrightarrow \Lambda'$ be a ring homomorphism (not necessarily commutative rings). Consider a functor $\mathcal{F} : \mathrm{mod}_\Lambda \longrightarrow \mathrm{mod}_{\Lambda'}$ given by $\mathcal{F}(M) = \Lambda' \otimes_\Lambda M$. Then $\mathcal{F}(M)$ is a Λ'-module. $\mathcal{F}$ also preserves properties of being free, projective or finitely generated, etc. Hence the correspondence $[P] \longmapsto [\mathcal{F}(P)]$ gives rise to a homomorphism $f_* : K_0\Lambda \longrightarrow K_0\Lambda'$. It is easy to see that K_0 is a covariant functor from the category of rings with identity to the category of abelian groups.

Here are some computations: Let $\mathrm{F}Ab$ and $\mathrm{F}gAb$ denote the categories of finite $\mathbb{Z}$-modules and finitely generated abelian $\mathbb{Z}$-modules respectively. Then $K_0\mathrm{F}Ab \cong \mathbb{Q}^+$ and $K_0\mathrm{F}gAb \cong \mathbb{Z}$. If Λ is a local ring or a principal ideal domain then $K_0\Lambda \cong \mathbb{Z}$ with generator $[\Lambda]$. In particular, if $\mathbb{F}_{p^\ell}$ is a finite field with p^ℓ elements then $K_0\mathbb{F}_{p^\ell} \cong \mathbb{Z}$.

III.2. $K_1\Lambda$

Let $M_n(\Lambda)$ denote the ring of n by n matrices with integers in an associative ring with 1. Let $GL_n(\Lambda)$ be the group of units of $M_n(\Lambda)$, i.e. the group of invertible matrices of $M_n(\Lambda)$. We will call $GL_n(\Lambda)$ the *general linear group*. A matrix which differs from the identity matrix by only an off diagonal element $\lambda \in \Lambda$ will be called *elementary* and will be denoted e_{ij}^λ, $i \neq j$. Since $\left(e_{ij}^\lambda\right)^{-1} = e_{ij}^{-\lambda}$ it is easy to check that multiplication of a matrix on the left or on the right by an elementary matrix corresponds to performing elementary row or column operations on such matrix. Also the reader can check that

$$\left[e_{ij}^\lambda, e_{k,\ell}^\mu\right] = \begin{cases} 1 & \text{if } j \neq k, \ i \neq \ell \\ e_{i\ell}^{\lambda\mu} & \text{if } j = k, \ i \neq \ell \\ e_{kj}^{-\lambda\mu} & \text{if } j \neq k, \ i = \ell \end{cases}$$

where $\left[e_{ij}^\lambda, e_{k,\ell}^\mu\right] = e_{ij}^\lambda e_{k\ell}^\mu \left(e_{ij}^\lambda\right)^{-1} \left(e_{k\ell}^\mu\right)^{-1}$, $n \geq 3$.

Let $E_n(\Lambda)$ denote the subgroup of $GL_n(\Lambda)$ generated by all $e_{ij}^\lambda, \lambda \in \Lambda$, $1 \leq i \neq j \leq n$, called the *elementary linear group* of Λ. If we identify each matrix $A \in GL_n(\Lambda)$ with the matrix $\begin{pmatrix} A & 0 \\ 0 & 1 \end{pmatrix} \in GL_{n+1}(\Lambda)$ we obtain inclusions $GL_1(\Lambda) \subset GL_2(\Lambda) \subset GL_3(\Lambda) \subset \ldots$. Lets denote by $GL(\Lambda)$ the direct limit of $GL_n(\Lambda)$ and call it the *stable* or *infinite general linear group of* Λ. The inclusion $GL_n(\Lambda) \hookrightarrow GL_{n+1}(\Lambda)$ restricts to the inclusion $E_n(\Lambda) \hookrightarrow E_{n+1}(\Lambda)$ and in $GL(\Lambda)$ the subgroup $E(\Lambda) = \varinjlim E_n(\Lambda)$ is called the *stable* or *infinite elementary group of* Λ.

Lemma 2.1. *(Whitehead)* $[GL(\Lambda), GL(\Lambda)] = E\Lambda$.

Proof. It is easy to see that every elementary matrix can be expressed as a commutator of other two elementary matrices for $n \geq 3$. Then $[E_n(\Lambda), E_n(\Lambda)] = E_n(\Lambda)$ and hence $[E(\Lambda), E(\Lambda)] = E(\Lambda) \subset GL(\Lambda)$. To see that $[GL(\Lambda), GL(\Lambda)] \subset E(\Lambda)$ let $A, B \in GL_n(\Lambda)$. Then, in $GL_{2n}(\Lambda)$

$$\begin{bmatrix} ABA^{-1}B^{-1} & 0 \\ 0 & I \end{bmatrix} = \begin{bmatrix} A & 0 \\ 0 & A^{-1} \end{bmatrix} \begin{bmatrix} B & 0 \\ 0 & B^{-1} \end{bmatrix} \begin{bmatrix} (BA)^{-1} & 0 \\ 0 & BA \end{bmatrix} \qquad (2.2)$$

Lets see that the commutator $ABA^{-1}B^{-1}$ in $GL_{2n}(\Lambda)$ can be expressed as the product of elementary matrices in $GL_{2n}(\Lambda)$. It is enough to prove that any matrix of 2.2 can be reduced to I_{2n} by means of elementary operations. Since

$$\begin{bmatrix} X & 0 \\ 0 & X^{-1} \end{bmatrix} = \begin{bmatrix} I & X \\ 0 & I \end{bmatrix} \begin{bmatrix} I & 0 \\ -X^{-1} & I \end{bmatrix} \begin{bmatrix} I & X \\ 0 & I \end{bmatrix} \begin{bmatrix} 0 & -I \\ I & 0 \end{bmatrix}$$

and since

$$\begin{bmatrix} I & X \\ 0 & I \end{bmatrix} \quad \text{and} \quad \begin{bmatrix} I & 0 \\ -X^{-1} & I \end{bmatrix}$$

can be reduced to I_{2n} using elementary row operations while

$$\begin{bmatrix} 0 & -I \\ I & 0 \end{bmatrix} \sim \begin{bmatrix} I & -I \\ I & 0 \end{bmatrix} \sim \begin{bmatrix} I & 0 \\ 0 & I \end{bmatrix}$$

by means of row and column operations respectively we have that $[GL_n(\Lambda), GL_n(\Lambda)] \subset E_{2n}(\Lambda)$, hence $[GL(\Lambda), GL(\Lambda)] \subset E(\Lambda)$. $\square$

As a consequence of Whitehead's lemma we have that $E(\Lambda)$ is a normal subgroup of $GL(\Lambda)$.

Definition 2.3. The quotient group $GL(\Lambda)/E(\Lambda)$ is called the *first algebraic K-theory group of the ring Λ* and is denoted by $K_1\Lambda$.

By lemma 2.1, $K_1\Lambda = GL(\Lambda)/[GL(\Lambda), GL(\Lambda)]$ and by II.2.3, $K_1\Lambda \cong H_1(GL(\Lambda), \mathbb{Z})$. Observe that K_1 is a functor from the category of rings to the category of abelian groups since a ring homomorphism $f : \Lambda \longrightarrow \Lambda'$ induces a group homomorphism $f_* : GL(\Lambda) \longrightarrow GL(\Lambda')$ that sends $E(\Lambda)$ to $E(\Lambda')$. Then f induces a group homomorphism $K_1(f) : K_1(\Lambda) \longrightarrow K_1(\Lambda')$ that satisfy the functorial properties for K_1.

Suppose that Λ is commutative. Then we can consider the determinant of a matrix as a homomorphism $\det : GL(\Lambda) \longrightarrow \Lambda^*$ where Λ^* denotes the units of Λ.

Let

$$SL(\Lambda) = \ker(\det) = \bigcup_{n=1}^{\infty} SL_n(\Lambda)$$

where

$$SL_n(\Lambda) = \ker(\det : GL_n(\Lambda) \longrightarrow \Lambda^*)$$

and call it the *stable or infinite special group of Λ*. It is clear that $E_n(\Lambda) \subset SL_n(\Lambda)$ for all n and $E(\Lambda) \subset SL(\Lambda)$. Observe that $\det : GL(\Lambda) \longrightarrow \Lambda^*$ induces a homomorphism also denoted by $\det : K_1(\Lambda) = GL(\Lambda)/E(\Lambda) \longrightarrow \Lambda^*$ that has an inverse $\Lambda^* = GL_1(\Lambda) \hookrightarrow GL(\Lambda) \longrightarrow K_1(\Lambda)$. If we define $SK_1(\Lambda) = SL(\Lambda)/E(\Lambda) = \ker(\det : K_1(\Lambda) \longrightarrow \Lambda^*)$ we obtain $K_1(\Lambda) = SK_1(\Lambda) \oplus \Lambda^*$, i.e. the short exact sequence

$$1 \longrightarrow SL(\Lambda)/E(\Lambda) \longrightarrow GL(\Lambda)/E(\Lambda) \overset{\det}{\longrightarrow} \Lambda^* \longrightarrow 1$$

splits. Since Λ^* can be considered known, the computation of $K_1(\Lambda)$ restricts to the computation of $SK_1(\Lambda)$. Observe that $SK_1(\Lambda)$ is trivial if and only if for any matrix $A \in SL_n(\Lambda)$ we can transform the matrix $\left[\begin{pmatrix} A & 0 \\ 0 & I_k \end{pmatrix}\right]$, for a suitable k, into the identity I_{n+k} by means of row and column operations. If $SK_1(\Lambda)$ is trivial then $K_1(\Lambda) \cong \Lambda^*$ and the homomorphism det is universal. This is the case if Λ is an integral domain, a local ring or a finite commutative ring. Then if $\mathbb{F}$ is a field, $K_1(\mathbb{F}) \cong \mathbb{F}^*$ and $K_1(\mathbb{F}[x]) \cong \mathbb{F}^*$; if $\Lambda = \mathbb{Z}$, $K_1(\mathbb{Z}) = \mathbb{Z}^* = \{\pm 1\}$.

There is a definition given by Bass for $K_1(_\Lambda P)$, which we will not describe here, such that there is a natural equivalence between $K_1(\Lambda)$ and $K_1(_\Lambda P)$ by looking at K_1 as a quotient of K_0 with an extra relation. See [B], [LL].

III.3. $K_2\Lambda$

Let us define a group by means of generators and relations that imitate the behavior of the elementary matrices introduced in section 2. Observe that the relations given in section 2 for the commutator $[e_{ij}^\lambda, e_{k\ell}^\mu]$ are not a set of relations that define $E(\Lambda)$.

Definition 3.1. For $n > 2$, define the *Steinberg group* $\mathrm{St}_n(\Lambda)$ as the non-abelian group given by the following presentation:

generators : x_{ij}^λ, $1 \le i \ne j \le n$, $\lambda \in \Lambda$

relations :   i) $x_{ij}^\lambda x_{ij}^\mu = x_{ij}^{\lambda+\mu}$

   ii) $[x_{ij}^\lambda, x_{k\ell}^\mu] = \begin{cases} 1 & j \ne k, i \ne \ell \\ x_{i\ell}^{\lambda\mu} & i \ne \ell, j = k \end{cases}$.

The relations (i) and (ii) are called *Steinberg relations* and obviously are similar to those of e_{ij}^λ of section 2. We can say that $\mathrm{St}_n(\Lambda)$ is the quotient group F/R where F is the free group generated by the symbols x_{ij}^λ and R is the smallest normal subgroup for which relations (i) and (ii) are valid. Note that for $n = 2$ the Steinberg relations are inadequate.

There is an epimorphism $\phi_n : \mathrm{St}_n(\Lambda) \longrightarrow E_n(\Lambda)$ given by $\phi_n\left(x_{ij}^\lambda\right) = e_{ij}^\lambda$. Denote $\mathrm{St}(\Lambda) = \varinjlim \mathrm{St}_n(\Lambda)$.

Definition 3.1. The second algebraic K-group of a ring Λ is defined as the kernel of the epimorphism ϕ. This means, $K_2(\Lambda) = \ker\big(\phi : \mathrm{St}(\Lambda) \longrightarrow E(\Lambda) \subset GL(\Lambda)\big)$.

We have the following exact sequences

$$1 \longrightarrow K_2(\Lambda) \longrightarrow \mathrm{St}(\Lambda) \overset{\phi}{\longrightarrow} E(\Lambda) \longrightarrow 1$$

and

$$1 \longrightarrow K_2(\Lambda) \longrightarrow \mathrm{St}(\Lambda) \overset{\phi}{\longrightarrow} GL(\Lambda) \longrightarrow K_1(\Lambda) \longrightarrow 1 \ .$$

It is clear that if $K_2\Lambda = 0$ then $\mathrm{St}(\Lambda) \cong E(\Lambda)$ and then the Steinberg relations form a set of relations that define $E(\Lambda)$. We can see $K_2\Lambda$ as a measure of the degree in which the Steinberg relations fail to be a set of relations that define $E(\Lambda)$ or as the set of non-trivial relations between elementary matrices.

If $f : \Lambda \longrightarrow \Lambda'$ is a ring homomorphism, it is immediate that f induce a group homomorphism $\mathrm{St}(f) : \mathrm{St}(\Lambda) \longrightarrow \mathrm{St}(\Lambda')$ given by $x_{ij}^{\lambda} \longmapsto x_{ij}^{f(\lambda)}$ that makes $\mathrm{St}(\_)$ a functor from the category of rings to the category of groups. It restricts to a homomorphism $K_2(f) : K_2(\Lambda) \longrightarrow K_2(\Lambda')$ making K_2 a covariant functor from the category of rings to the category of groups. It can be proved that for any ring Λ, $Z\big(E(\Lambda)\big) = 1$ ($Z(\ )$ denotes the center) and $Z\big(\mathrm{St}(\Lambda)\big) = K_2(\Lambda)$ making K_2 a covariant functor into the category of abelian groups.

We have the following result

Theorem 3.2. *Let $\rho : G \longrightarrow E(\Lambda)$ be a central extension. Then there is a unique homomorphism $\psi : \mathrm{St}(\Lambda) \longrightarrow G$ such that the following diagram commutes*

$$
\begin{array}{ccc}
\mathrm{St}(\Lambda) & \overset{\varphi}{\longrightarrow} & E(\Lambda) \\
\psi \downarrow & & \| \\
G & \overset{\rho}{\longrightarrow} & E(\Lambda)
\end{array}
$$

Since $\phi : \mathrm{St}(\Lambda) \longrightarrow E(\Lambda)$ is a universal central extension whose kernel is $K_2\Lambda$, as we saw in II.3, $K_2\Lambda \cong H_2(E(\Lambda), \mathbb{Z})$. This relates H_2 with K_2.

We will give without proof the following

Theorem 3.3. *Let Λ and Λ' be rings. Then*

$$\mathrm{St}(\Lambda \times \Lambda') \cong \mathrm{St}(\Lambda) \oplus \mathrm{St}(\Lambda')$$

and

$$K_2(\Lambda \times \Lambda') \cong K_2(\Lambda) \oplus K_2(\Lambda') \ .$$

Finally, some computations of $K_2\Lambda$ for diverse rings Λ are:
(a) If $\Lambda = \mathbb{F}_{p^\ell}$ is a finite field with p^ℓ elements (p a prime number) then $K_2(\mathbb{F}_{p^\ell}) = 1$.
(b) If $\Lambda = \mathbb{Z}$ then $K_2(\mathbb{Z}) \cong \mathbb{Z}/2$.

Bibliography and References

[B]   Bass, H:: Algebraic K-Theory. Benjamin, 1968.

Brown, K.S.: Cohomology of Groups. Grad. Texts in Math. **87**. Springer Verlag, 1982.

Cartan, H:, Eilenberg, S.: Homological Algebra. Princeton University Press.

Hilton, P., Stammbach, U.: A Course in Homological Algebra. Grad. Texts in Math. **4**. Springer Verlag , 1971.

Lam, T.Y., Siu, M.K.: K_0 and K_1 – An Introduction to Algebraic K-Theory. Amer. Math. Monthly, April, 1975, p. 329–364.

[LL] Lluis-Puebla, E.: Algebra Homológica, Cohomología de Grupos y K-Teoría Algebraica Clásica. Addison Wesley Iberoamericana, 1990.

Mac Lane, S.: Homology. Springer Verlag, 1975.

Milnor, J.R.: Introduction to Algebraic K-Theory. Annals of Math. Studies **72**. Princeton, 1971.

Rotman J.: An Introduction to Homological Algebra. Academic Press, 1979.

Silvester, J.R.: Introduction to Algebraic K-Theory. Chapman and Hall, 1981.

Introduction to Algebraic K-Theory and Cyclic Homology

Jean-Louis Loday

Institut de Recherche Mathématique Avancée, ULP et CNRS, 7 rue R. Descartes, F-67084 Strasbourg, France

The study of the general linear group of an arbitrary ring has given rise to algebraic K-theory K_*. Similarly the study of the Lie algebras of matrices leads to what should be called *additive algebraic K-theory* K_*^+. In fact, we are, in a certain sense, able to compute K_*^+, that is, to reduce it to a simpler theory called *cyclic homology HC_**.

The analogy between K_* and K_*^+ is illustrated by the following equalities which can be taken as definitions in the rational case:

$$K_n(A) \otimes \mathbb{Q} = \mathrm{Prim}\, H_n(GL(A), \mathbb{Q}) \ ,$$
$$K_n^+(A) \otimes \mathbb{Q} = \mathrm{Prim}\, H_n(\mathfrak{gl}(A), \mathbb{Q}) \ .$$

Here GL is the general linear group, $\mathfrak{gl}$ the Lie algebra of matrices, and Prim denotes the primitive part.

There is a whole dictionary of the objects that we will encounter:

K-theory	G_m	$x + y - xy$	GL	SL	St	det	K_n	K_n^M
Cyclic homology	G_a	$x + y$	$\mathfrak{gl}$	sl	st	trace	$K_n^+ = HC_{n-1}$	$\Omega^{n-1}/d\Omega^{n-2}$

In fact the equality $K_n^+ = HC_{n-1}$ is a theorem that we will prove in §2.

Both of these theories are now intensively studied because they are connected to many parts of mathematics, like number theory (Galois theory, ζ-functions), differential geometry (pseudo-isotopies), mathematical physics (gauge groups, anomalies), combinatorics (Macdonald conjectures).

These are the notes written in 1985 from lectures given for the "Seminario de K-teoría algebraica" held at "Universidad Nacional Autónoma de México"on the occasion of its 75th birthday. A more comprehensive and recent version will be found in J.-L. Loday, "Cyclic Homology", to appear in the Springer series "Grundlehren der mathematischen Wissenschaften".

I thank all the participating mathematicians and particularly Professor Emilio Lluis-Puebla for the excellence of the organization.

Contents

1. Hochschild Homology and Rational Cyclic Homology

Let k be a commutative ring and A a k-algebra (associative but not necessarily commutative). We denote by A^n instead of $A^{\otimes n}$ the tensor product of n copies of A over k. Let $b : A^{n+1} \longrightarrow A^n$ be the Hochschild boundary

$$b\left(a_0, \ldots, a_n\right) = \sum_{i=0}^{n-1} (-1)^i \left(a_0, \ldots, a_i a_{i+1}, \ldots, a_n\right)$$
$$+ (-1)^n \left(a_n a_0, a_1, \ldots, a_{n-1}\right) \ .$$

It is an immediate verification that $b^2 = 0$. Therefore we get a *Hochschild complex*

$$\cdots \longrightarrow A^{n+1} \longrightarrow A^n \longrightarrow \ldots \longrightarrow A$$

and *Hochschild homology groups* denoted $H_n(A, A)$. First computation: $H_0(A, A) = A/[A, A]$ where $[A, A]$ is generated by additive commutators $ab - ba$. We have also $H_n(k, k) = k$ if $n = 0$ and 0 otherwise. If A is commutative $H_1(A, A) = \Omega^1_{A/k}$ (Absolute Kähler differentials).

Consider now the operator $t : A^{n+1} \longrightarrow A^{n+1}$

$$t\left(a_0, \ldots, a_n\right) = (-1)^n \left(a_n, a_0, \ldots, a_{n-1}\right)$$

(cyclic operator of order $n + 1$) and define

$$C_n^\lambda(A) = A^{n+1}/(1 - t) \ .$$

In other words the action is made trivial.

Lemma 1.1. *The boundary b is well defined on $C_n^\lambda(A)$.*

Proof. This is a straightforward computation which can be seen as follows: $b = \sum_{i=0}^n (-1)^i d_i$ and $d_i t = t d_{i-1}$ for $1 \leq i \leq n$, $d_0 t = d_n$. Then put $b' = \sum_{i=0}^{n-1} (-1)^i d_i$, it follows that $b(1 - t) = (1 - t)b'$ which proves the lemma. $\square$

As a consequence we get *Connes' complex*

$$C_*^\lambda(A) \ldots \longrightarrow C_n^\lambda(A) \xrightarrow{\ b\ } C_{n-1}^\lambda(A) \longrightarrow \ldots \longrightarrow C_0^\lambda(A) \ .$$

Definition 1.2. If $k \supset \mathbb{Q}$ we define *cyclic homology* as

$$HC_n(A) = H_n(C_*^\lambda(A), b) \ .$$

Examples.

$$HC_0(A) = A/[A,A] \ ,$$
$$HC_1(A) = \Omega^1_{A/k}/dA \quad \text{if } A \text{ is commutative} \ ,$$
$$HC_n(k) = \begin{cases} k & \text{if } n \text{ is even} \\ 0 & \text{if } n \text{ is odd} \end{cases} \ .$$

Now we can state the main theorem.

Theorem 1.3. ([L-Q], [T]). *Let k be a commutative ring with $k \supset \mathbb{Q}$ and A an associative k-algebra. Then*

$$K_n^+(A) \cong HC_{n-1}(A) \ .$$

In other words

$$\operatorname{Prim} H_n(g\ell(A), k) = HC_{n-1}(A)$$

or

$$H_*(g\ell(A), k) = \Lambda^* HC_{*-1}(A)$$

(graded exterior algebra).

So we get in low dimensions:
$H_1(g\ell(A), k) = HC_0(A)$ which is obviously true integrally,
$H_2(sl(A), k) = HC_1(A)$, proved integrally in [K-L].

In this last chapter it is also proved that $sl(A)$ has a universal central extension denoted $st(A)$ and by the above theorem we have

$$H_3(st(A), k) = HC_2(A) \ .$$

2. The Proof of $K_n^+ = HC_{n-1}$

We prove that $\operatorname{Prim} H_*(g\ell(A), k) = HC_{*-1}(A)$ as follows. Let S_n denote the permutation group on n letters.

a. Consider $\Lambda^n g\ell_p(A) = (g\ell_p(A)^{\otimes n}) \otimes_{S_n} (\operatorname{sgn})$, sgn = signature representation. As a k-vector space $g\ell_p(A) = g\ell_p(k) \otimes A$.

Notation: $\mathbf{g}_p = g\ell_p(k)$ and $\mathbf{g} = g\ell(k) = \lim_p g\ell_p(k)$. We have

$$\Lambda^n g\ell(A) = (\mathbf{g}^{\otimes n} \otimes A^{\otimes n}) \otimes_{S_n} (\operatorname{sgn}) \ .$$

b. Consider the Koszul complex $\ldots \longrightarrow \Lambda^n g\ell_p(A) \longrightarrow \Lambda^{n-1} g\ell_p(A) \longrightarrow \ldots$ whose homology is $H_n(g\ell(A), k)$.

Proposition 2.1. *The homology of this complex is isomorphic to the homology of the complex of coinvariants*

$$\ldots \longrightarrow \left(\Lambda^n g\ell_p(A)\right)_{g_p} \longrightarrow \left(\Lambda^{n-1} g\ell_p(A)\right)_{g_p} \longrightarrow \ldots .$$

Proof. g_p is acting on $g_p(A)$ by $u \cdot \alpha = [u, \alpha]$ so is acting on $g\ell_p(A)^{\otimes n}$ by $u \cdot (\alpha_1 \otimes \ldots \otimes \alpha_n) = \sum_1^n \alpha_1 \otimes \ldots \otimes u \cdot \alpha_i \otimes \otimes \alpha_n$.

Coinvariants means that one makes this action trivial: $[u, -] = 0$. Main point: $g\ell_p(k)$ is reductive (because k is a car. 0 field) which means that any $g\ell_p$-module is the sum of simple modules (we look at the modules $\Lambda^n g\ell_p(A)$). In $H_*(g\ell(A), k)$, $g\ell_p(k)$ is acting trivially and the homology is made of some of the simple modules (because d is equivariant). Therefore if we split $\Lambda^* g\ell_p(A)$ into trivial modules $\oplus$ non trivial modules then the sum of the trivial modules is an acyclic sub-complex and the sum of the trivial ones is $(\Lambda^n g\ell_p(A))_{g\ell_p(k)}$. Whence the proposition. □

c. From a. and b. we see that we need to compute

$$L_n = \left(\Lambda^n g\ell_p(A)\right)_{g_p} = \left(\left(g_p^{\otimes n} \otimes A^{\otimes n}\right) \otimes_{S_n} (\text{sgn})\right)_{g_p}$$

and d on it.

Note that g_p is acting only on $g_p^{\otimes n}$ therefore we have

$$L_n = \left(\left(g_p^{\otimes n}\right)_{g_p} \otimes A^{\otimes n}\right) \otimes_{S_n} (\text{sgn}) .$$

Proposition 2.2. (Invariant Theory). $\left(g_p^{\otimes n}\right)_{g_p} \xrightarrow{\sim} k[S_n]$ *if $p \geq n$ and the action of S_n on $g_p^{\otimes n}$ (permutation of variables) translates to conjugation on $k[S_n]$.*

We refer to [W] or [D-P] for the proof. However we make this isomorphism explicit. Let $M_1, \ldots, M_n$ be $p \times p$-matrices. The image of $M_1 \otimes \ldots \otimes M_n$ in $k[S_n]$ is $\sum a(\sigma)\sigma$ where $a(\sigma)$ is computed as follows. Write σ as a product of cycles $\sigma = (i_1 \ldots i_k)(j_1 \ldots j_\ell) \ldots$, then

$$a(\sigma) = \text{Tr}\,(M_{i_1} \ldots M_{i_k})\,\text{Tr}\,(M_{j_1} \ldots M_{j_\ell}) \ldots .$$

As a consequence of this proposition we have $L_n = ([S_n] \otimes A^{\otimes n}) \otimes_{Sn} (\text{sgn})$.

Remark. The matrices have disappeared.

d. Remind that we want to compute the primitive part of H_*. But we are dealing with Hopf algebras over a car. 0 field.

We apply the following proposition to $\mathcal{H} = (\oplus_n L_n, d)$.

Proposition 2.3. *Let $\mathcal{H}$ be a differential graded Hopf algebra over a characteristic zero field. If $\mathrm{Prim}\,\mathcal{H}$ is an abelian Lie algebra then*

$$H_*(\mathrm{Prim}\,\mathcal{H}, d) = \mathrm{Prim}\big(H_*(\mathcal{H}, d)\big) \ .$$

Proof. By the Cartan-Milnor-Moore theorem we have $\Lambda^*\mathrm{Prim}\,\mathcal{H} = \mathcal{H}$ so it suffices to prove that

$$\Lambda^* H_*(P, d) = H_*(\Lambda^* P, d) \quad \text{where} \quad P = \mathrm{Prim}\,\mathcal{H} \ .$$

As we are working over a field, the complex (P, d) can be written $(P, d) = (P', d') \oplus \big(H(P), 0\big)$ where (P', d') is acyclic. It is immediate that the formula above is true if $d = 0$, therefore it suffices to prove it for acyclic complexes. Splitting an acyclic complex into pieces it suffices to prove the formula for

$$X_* : \ldots \longrightarrow 0 \longrightarrow X \xrightarrow{id} X \longrightarrow 0 \longrightarrow \ldots \ .$$

Then $\Lambda^*(X_*)$ splits into the sum of the complexes $\Lambda^r X_*$, $r > 0$ which are acyclic.
$\square$

The problem now is to compute $\mathrm{Prim}\,(\oplus_n L_n, d)$. Remember that Prim comes from the coproduct structure: $\mathbf{g} \xrightarrow{\Delta} \mathbf{g} \times \mathbf{g}$. How does the isomorphism $(\mathbf{g}^{\otimes n})_{\mathbf{g}} \xrightarrow{\sim} k[S_n]$ behaves with it? The answer is

$$\Delta : k[S_n] \longrightarrow \oplus_{a+b=n} k[S_a] \otimes k[S_b]$$
$$\sigma \longrightarrow \Sigma\, \sigma_I \otimes \sigma_J$$

where the sum is over all partitions (I, J) of $\{1, 2, \ldots, n\}$ such that $\sigma(I) \subset I$ and $\sigma(J) \subset J$. (It is understood that if $i_1 < i_2 < \ldots < i_a$ are the integers of I we use the bijection $\ell \mapsto i_\ell$ to identify S_a to a subgroup of S_n).

When is σ primitive? We want the partition to be $\phi \bigcup \{1, 2, \ldots, n\}$. Then σ has to have only one cycle of length n (i.e. conjugate to $\tau = (1, 2, \ldots, n)$) and $\Delta(\sigma) = 1 \otimes \sigma + \sigma \otimes 1$.

Notation: U_n = conjugation class of τ in S_n.

Proposition 2.4.

$$\mathrm{Prim}\,(\oplus L_n) = \oplus_n \underbrace{\left(k[U_n] \otimes A^{\otimes n}\right) \otimes_{S_n} (\mathrm{sgn})}_{\mathrm{Prim}\,L_n}$$

in other words $k[U_n]$ is induced from the trivial module k over $\mathbb{Z}/n\mathbb{Z}$ by the map

$$\mathbb{Z}/n\mathbb{Z} \longrightarrow S_n,\ 1 \longmapsto \tau\ .$$

Hence

$$\begin{aligned}
\mathrm{Prim}\,L_n &= \left(k\left[S_n/(\mathbb{Z}/n\mathbb{Z})\right] \otimes A^{\otimes n}\right) \otimes_{S_n} (\mathrm{sgn}) \\
&= \left(k \otimes A^{\otimes n}\right) \otimes_{\mathbb{Z}/n\mathbb{Z}} (\mathrm{sgn}) = C^{\lambda}_{n-1}(A)\ .
\end{aligned}$$

e. It suffices now to see what happens to d. We consider the diagram

$$\Lambda^n g\ell(A) \xrightarrow{\quad d \quad} \Lambda^{n-1} g\ell(A)$$

$$(\Lambda^n g\ell(A))_{g\ell} \qquad\qquad .$$

$$\wr\wr \qquad\qquad\qquad\qquad \wr\wr$$

$$\left(k[S_n] \otimes A^{\otimes n}\right) \otimes_{S_n} (sgn) \qquad .$$

$$\left(k[U_n] \otimes A^{\otimes n}\right) \otimes_{S_n} (sgn) \qquad .$$

$$\wr\wr \qquad\qquad\qquad\qquad \wr\wr$$

$$C_n(A) \dashrightarrow^{\ ?\ } C_{n-1}(A)$$

We follow the following element

$$E_{12}^{a_1} \wedge E_{23}^{a_2} \wedge \ldots \wedge E_{n1}^{a_n} \quad \longmapsto \quad \sum_1^n \pm E_{12}^{a_1} \wedge \ldots \wedge E_{i\ i+2}^{a_i a_{i+1}} \wedge \ldots \wedge E_{n1}^{a_n}$$

idem

use part c, that is

$$\mathrm{Tr}(E_{12}^1 E_{23}^1 \ldots E_{n1}^1) = 1.$$

$\tau \otimes (a_1, \ldots, a_n)$

idem

$\|$

$$(a_1, \ldots, a_n) \quad \longmapsto \quad b(a_1, \ldots, a_n)$$

This ends the proof of the theorem.   □

3. Integral Cyclic Homology [L-Q]

Let K be a commutative ring and A a (not necessarily commutative) associative algebra.

On $A^{\otimes n+1}$ we have the following operators:

$$d_i : A^{\otimes n+1} \longrightarrow A^{\otimes n}, (a_0, \ldots, a_n) \longmapsto (a_0, \ldots, a_i a_{i+1}, \ldots, a_n)$$
$$\text{for} \quad 0 \le i < n$$
$$d_n : A^{\otimes n+1} \longrightarrow A^{\otimes n}, (a_0, \ldots, a_n) \longmapsto (a_n a_0, a_1, \ldots, a_{n-1})$$
$$t : A^{\otimes n+1} \longrightarrow A^{\otimes n+1}, (a_0, \ldots, a_n) \longmapsto (-1)^n (a_n, a_0, \ldots, a_{n-1})$$
$$\text{(cyclic permutation) .}$$

Put $b = \sum_{i=0}^n (-1)^i d_i$, $b' = \sum_{i=0}^{n-1} (-1)^i d_i$, $N = 1 + t + \ldots + t^n$. We abbreviate $A^{\otimes n}$ in A^n.

As A^{n+1} inherits a $\mathbb{Z}/n+1$-action from the operator t, then there is a very simple resolution using $(1-t)$ and N. This permits us to construct a diagram $C_{**}(A)$

$$
\begin{array}{ccccccc}
\downarrow & & \downarrow & & \downarrow & & \downarrow \\
A^{n+1} \xleftarrow{1-t} & & A^{n+1} \xleftarrow{N} & & A^{n+1} \xleftarrow{1-t} & & A^{n+1} \longleftarrow \\
b \downarrow & -b' \downarrow & & n \downarrow & & -b' \downarrow \\
A^n \longleftarrow & & A^n \longleftarrow & & A^n \longleftarrow & & A^n \longleftarrow \\
\downarrow & & \downarrow & & \downarrow & & \downarrow \\
\cdots & & \cdots & & \cdots & & \cdots
\end{array}
$$

Proposition 3.1. *This diagram is a bicomplex, i.e.*

$$b(1-t) - (1-t)b' = 0 \quad \text{and} \quad -b'N + Nb = 0 .$$

Proof. It is more or less a straightforward computation (cf. [L-Q]). Note that in these formulas t represents either the generator of $\mathbb{Z}/n+1$ or of the generator of $\mathbb{Z}/n$ depending on its place in the formula. □

Definition 3.2. *Cyclic homology of A over k is defined by*

$$HC_n(A) = H_n(\mathrm{Tot}\, C_{**}(A)) .$$

Here Tot means the total complex. The grading is such that A^{n+1} in the zero-th column is of degree n.
So $HC_0(A) = A/[A, A]$.

Proposition 3.3. *If $k \supset \mathbb{Q}$ then $HC_n(A) = H_n(C_*(A), b)$ where $C_n(A) = A^{\otimes n+1}/(1-t)$.*

So our new definition agrees with the former one in characteristic zero.

Proof. Homology of a row is $H_p(\mathbb{Z}/n, A^{\otimes n})$ which is trivial if $p > 0$ and $A^{\otimes n}/(1-t)$ if $p = 0$; whence the result. □

One of the main properties of cyclic homology which was first proved by Connes [C] with the first definition is the following periodicity exact sequence for which we give the proof of [L-Q].

Theorem 3.4. *There is a long exact sequence*

$$\cdots \longrightarrow H_n(A, A) \xrightarrow{I} HC_n(A) \xrightarrow{S} HC_{n-2}(A) \xrightarrow{B} H_{n-1}(A, A) \longrightarrow \cdots .$$

Proof. Consider the following exact sequence of bicomplexes
$$0 \longrightarrow (\text{first 2 columns}) \hookrightarrow C_{**}(A) \twoheadrightarrow C_{**}(A)[-2] \longrightarrow 0.$$

In order to understand the homology of the first bicomplex we remark that the b' column is acyclic, therefore the homology is $H_*(A, A)$. The homology exact sequence associated to an exact sequence of bicomplexes gives the result.   □

In order to compare cyclic homology to de Rham homology one needs to modify $C_{**}(A)$ slightly by removing the acyclic columns. For this purpose we state the "removing acyclic complexes lemma".

Lemma 3.5. *Let* $\ldots \longrightarrow A_{i+1} \oplus B_{i+1} \longrightarrow A_i \otimes B_i \longrightarrow \ldots$ *be a complex such that the differential* $d = \left(\begin{smallmatrix} \alpha & \beta \\ \gamma & \delta \end{smallmatrix}\right)$ *has the following property:* (B_*, δ) *is a complex and is homotopy equivalent to zero, i.e.* $\exists s : B_i \longrightarrow B_{i+1}$ *such that* $\delta s + s\delta = id$.

Then the complex $(A_* B_*, d)$ *is quasi-isomorphic to* $(A_*, \alpha - \beta s\gamma)$.

Proof. The map $(id, -s\gamma) : A_i \longrightarrow A_i \oplus B_i$ is an injection of complexes (with differential $\alpha - \beta s\gamma$ for A_*) and its cokernel is the acyclic complex (B_*, δ_*).   □

Apply this lemma to the b' columns of $C_{**}(A)$. We get a new bicomplex

$$
\begin{array}{ccccccc}
b \downarrow & & b \downarrow & & b \downarrow & \\
A^3 & \xleftarrow{B} & A^2 & \xleftarrow{B} & A & \\
b \downarrow & & b \downarrow & & \\
A^2 & \xleftarrow{B} & A & & \\
b \downarrow & & & & \\
A & & & &
\end{array}
$$

where $B = (1 - t)sN$ and $s(a_0, \ldots, a_n) = (1, a_0, \ldots, a_n)$.

Suppose now that A is commutative.

Let Ω_A^1 be the A-module of Kähler differentials and let $\Lambda_A^n \Omega_A^1 = \Omega_A^n$ be the A-module of n-differentials. If $k \supset \mathbb{Q}$ there is a well-defined map $\gamma : A^n \longrightarrow \Omega_A^n$ given by $(a_0, \ldots, a_n) \longmapsto \frac{1}{n!} a_0 da_1 \ldots da_n$. One checks that $\gamma \circ b = 0$ and $\gamma \circ B = d \circ \gamma$ where $d : \Omega_A^n \longrightarrow \Omega_A^{n+1}$ is the classic differentiation operator. Therefore γ defines a morphism of complexes from $B_{**}(A)$ to

$$\downarrow \qquad\qquad \downarrow \qquad \downarrow$$

$$\Omega_A^2 \quad \xleftarrow{d} \quad \Omega_A^1 \quad \xleftarrow{d} \quad \Omega_A^0$$

$$0 \downarrow \qquad\quad 0 \downarrow$$

$$\Omega_A^1 \quad \xleftarrow{d} \quad \Omega_A^0$$

$$0 \downarrow$$

$$\Omega_A^0$$

The n-th homology group of the total complex of this bicomplex is

$$\Omega_A^n/d\Omega_A^{n-1} \oplus H_{DR}^{n-2}(A) \oplus H_{DR}^{-4}(A) \oplus \dots$$

where $H_{DR}^*(A)$ stands for de Rham cohomology of A.

When A is a smooth ring then a Theorem of Hochschild-Kostant and Rosenberg asserts that the map $H_n(A, A) \longrightarrow \Omega_A^n$ is an isomorphism. As a consequence we deduce

Theorem 3.6. *If $k \supset \mathbb{Q}$ and if A is commutative and smooth the map γ induces an isomorphism*

$$HC_n(A) \cong \Omega_A^n/d\Omega_A^{n-1} \oplus H_{DR}^{n-2}(A) \oplus H_{DR}^{n-4}(A) \oplus \dots \ .$$

Proof. The E^2-terms of the spectral sequences deduced from the bicomplexes are isomorphic by Hochschild-Kostant-Rosenberg. Thus so are the abutments. □

Remark. This theorem shows that cyclic (co)homology is a good candidate to play the role of de Rham cohomology when the ring is not commutative. This is in fact the case and was the motivation of Connes for inventing cyclic cohomology.

Operations in Cyclic Homology

We will particularly study two of them: the product and the action of derivations.

Suppose that A is commutative, then on $\oplus A^{n+1}$ (remind that A^n means $A^{\otimes n}$) there is the so-called shuffle product: if $x = (a, a_1, \dots, a_n) \in A^{n+1}$ and $x' = (a', a_{n+1}, \dots, a_{p+n}) \in A^{p+1}$ then

$$x \cdot y = \sum (aa', a_{\sigma(1)}, \dots, a_{\sigma(n+p)})$$

where the sum is over all permutations σ of $\{1, \dots, n+p\}$ such that $\sigma 1 < \sigma 2 < \dots < \sigma n$ and $\sigma(n+1) < \dots < \sigma(n+p)$.

It is well-known that this product induces a product in Hochschild homology.

For simplicity we define the product on cyclic homology by using the first definition (with $C_n^\lambda(A) = A^{n+1}/(1-t)$) though it is defined with the integral version (cf. [L-Q]).

First we describe B on the complex $C_*(A)^{\text{red}}$. This complex is $C_*^\lambda(A)$ divided by the subcomplex generated by the elements of the form $(a_0, a_1, \ldots, a_n)$ where $a_i = 1$ for some i, $1 \le i \le n$ (not $i = 0$!). The quotient complex has the same homology as the former one and we have the following:

Lemma 3.7. *On* $C_n(A)^{\text{red}}$ *the operator* B *is*

$$B(a_0, \ldots, a_n) = \sum_{i=0}^{n} (-1)^{in} (1, a_i, a_{i+1}, \ldots, a_n, a_0, \ldots, a_{i-1}) \ .$$

Definition . On $C_*(A)^{\text{red}}$ we define the product $*$ by the following formula

$$x \in C_n(A)^{\text{red}}, y \in C_p(A)^{\text{red}}, x * y = x \cdot B(y) \in C_{n+p+1}(A)^{\text{red}}$$

where the dot means shuffle product as defined above.

Proposition 3.8. *This* $*$-*product induces a product* $HC_n(A) \times HC_p(A) \longrightarrow HC_{n+p+1}(A)$ *which is graded commutative.*

We refer to [L-Q]§3 for the proof.

Examples. If $n = 1$, $HC_1(A) = \Omega_A^1/dA$ and $HC_0(A) \times HC_0(A) \longrightarrow HC_1(A)$ is given by $(a, b) \longmapsto adb$. More generally the product of $n + 1$ elements of $A = HC_0(A)$ factors through $\Omega_A^n/d\Omega_A^{n-1}$:

$$A \times \ldots \times A \longrightarrow \Omega_A^n/d\Omega_A^{n-1} \longrightarrow HC_n(A)$$
$$(a_0, \ldots, a_n) \longmapsto a_0 da_1 \ldots da_n \longmapsto a_0 * a_1 * \ldots * a_n \ .$$

It is easy to see that if $k \supset \mathbb{Q}$ the map $A^{n+1} \longrightarrow \Omega_A^n/d\Omega_A^{n-1}$, $(a_0, \ldots, a_n) \longmapsto \frac{1}{(n-1)!} a_0 da_1 \ldots da_n$ induces a map $HC_n(A) \longrightarrow \Omega_A^n/d\Omega_A^{n-1}$.

Proposition 3.9. *The composition*

$$\Omega_A^n/d\Omega_A^{n-1} \longrightarrow HC_n(A) \longrightarrow \Omega_A^n/\Omega_A^{n-1}$$

is multiplication by $n!$.

When A is smooth we can describe this product completely.

Proposition 3.10. *If $k \supset \mathbb{Q}$ and A is commutative and smooth the product on $HC_n = \Omega^n/d\Omega^{n-1} \oplus H_{DR}^{n-2} \oplus H_{DR}^{n-4} \oplus \ldots$ is given by*

$$x * y = \bar{x} \wedge d\bar{y}$$

where $\bar{x}$ (resp. $\bar{y}$) is the component in $\Omega^n/d\Omega^{n-1}$ (resp. $\Omega^p/d\Omega^{p-1}$) of $x \in HC_n$.

This is more easily proved using the product defined on integral cyclic homology so we refer to [L-Q] for the proof.

Action of a Derivation

Let A be commutative and D be a derivation, i.e.

$$D(\lambda a + \mu b) = \lambda Da + \mu Db, \quad \lambda, \mu \in k, \; a, b \in a \; ,$$
$$D(ab) = aDb + bDa \; .$$

D induces an operation on A^{n+1} still denoted D:

$$D(a_0, \ldots, a_n) = \sum_{i=0}^{n} (a_0, \ldots, Da_i, \ldots, a_n) \; .$$

This operation passes to the quotient to give on the homology

$$D_* : HC_n(\Lambda) \longrightarrow HC_n(\Lambda) \; .$$

Theorem 3.11. (Goodwillie [G]). *If D is a derivation of the commutative ring A then $S \circ D_* : HC_n(A) \longrightarrow HC_{n-2}(A)$ is trivial.*

We will not prove this theorem here. As we will see later it has some nice consequences as far as computation is concerned.

4. Cyclic Homology of Truncated Polynomial Rings

Let $A = k[t]/t^{\ell+1}$ be a truncated polynomial ring over a characteristic zero field k. If $\ell = 1$ it is usually called the *ring of dual numbers*.

First we compute its Hochschild homology $H_n(A, A)$.

Proposition 4.1. *If $A = k[t]/t^{\ell+1}$ then $H_n(A, A)$ is a k-vector space of dim ℓ when $n > 0$.*

Proof. Hochschild homology of A with coefficients in a bimodule M can be defined by $H_n(A, M) = \mathrm{Tor}_n^{A \times A^\circ}(A, M)$ where A° is the opposite ring. Taking the standard Hochschild resolution

$$\ldots \xrightarrow{b'} A^3 \xrightarrow{b'} A^2 \xrightarrow{b'} A$$

of A over $A \otimes A$ gives the Hochschild complex we were using previously.

However, in our particular case we can produce the following ad hoc resolution. Let $Q = k[x,y]/x^{\ell+1} = y^{\ell+1} = 0$ be an $A \otimes A$-module where the first factor acts by multiplication by x and the second one by multiplication by y (so $Q \simeq A \otimes A$). Then put $P(x,y) = x^{\ell} + x^{\ell-1}y + \dots xy^{\ell}$ so that $P(x,y)(x-y) = x^{\ell+1} - y^{\ell+1} = 0$.

The following is a free resolution of A:

$$\dots \xrightarrow{P(x,y)} Q \xrightarrow{x-y} Q \xrightarrow{P(x,y)} Q \xrightarrow{x-y} Q \twoheadrightarrow A$$

(compare with the free resolution of $\mathbb{Z}[\mathbb{Z}/\ell]$).

Then $H_n(A,A)$ is given by the homology of

$$\dots \longrightarrow A \xrightarrow{(\ell+1)t^{\ell}} A \xrightarrow{0} A \xrightarrow{(\ell+1)t^{\ell}} A \xrightarrow{0} A \ .$$

Therefore $H_0(A,A) = A$ (as we knew), $H_{2n+1}(A,A) \cong A/t^{\ell}$ and $H_{2n}(A,A) \cong A/k$ (if $n > 0$).

In order to compute $HC_n(A)$ we introduce the notion of *weight* on $A : w(t) = 1$. It extends to A^n by $w(t^{i_0}, \dots, t^{i_n}) = \sum_{i=0}^{n} i_j$. It is immediate to see that b and the cyclic operator respect the weight, so $HC_n(A) = \oplus_{w \geq 0} HC_{n,w}(A)$.

Define a derivation D on $A = k[t]/t^{\ell+1}$ by $Dt = t$. So D also respects the weight. In particular on $HC_{n,w}(A)$ we have $D_* = w \times id$. As a consequence D_* is trivial on $HC_{n,w}(A)$ as far as $w \neq 0$.

We apply now Goodwillie's result (theorem 3.11) to conclude that $S = 0$ on $HC_{n,w}$ for $w \neq 0$. Let $\widetilde{HC}_n(A) = HC_n(A)/HC_n(k)$ (here $HC_n(k)$ identifies to $HC_{n,0}(A)$).

Then the periodicity exact sequence reads as follows:

$$\dots \longrightarrow \tilde{H}_n(A,A) \longrightarrow \widetilde{HC}_n(A) \xrightarrow{0} \widetilde{HC}_{n-2}(A) \longrightarrow \tilde{H}_{n-1}(A,A) \longrightarrow \dots$$
$$\xrightarrow{0} \widetilde{HC}_0(A) \longrightarrow \tilde{H}_1(A,A) \longrightarrow \widetilde{HC}_1(A) \longrightarrow 0 \ .$$

As $\widetilde{HC}_0(A)$ and $\tilde{H}_1(A,A)$ are of dimension ℓ it follows that $\widetilde{HC}_1(A) = 0$. Finally $HC_{2n}(A)$ is of dimension $\ell + 1$ and $HC_{2n+1}(A) = 0$.

Proposition 4.2. *If k is a characteristic zero field then $HC_{2n+1}\left(k[t]/t^{\ell+1}\right) = 0$ and $HC_{2n}\left(k[t]/t^{\ell+1}\right)$ is of dimension $\ell + 1$ over k.*

$\square$

5. Quillen's Plus Construction and K_n ([Q], [L1])

This construction performed on a space with perfect fundamental group keeps the homology the same but kills the fundamental group.

A group G is called *perfect* if it is equal to its commutator group $[G,G]$. In other words the abelianized group $G_{ab} = G/[G,G]$ is trivial.

Theorem 5.1. ([Q], [L1]). *Let X be a connected cellular complex with base-point. Suppose that its fundamental group $\pi_1(X)$ is perfect. Then there exists a connected cellular space X^+ and an inclusion $i : X \longrightarrow X^+$ such that*

a) $\pi_1(X^+) = 1$,
b) i induces an isomorphism in homology

$$i_* : H_*(X, \mathbb{Z}) \xrightarrow{\sim} H_*(X^+, \mathbb{Z}) \ .$$

Proof. Let $f_\alpha : S^1 \longrightarrow X$, $\alpha \in A$, be a set of (base-point preserving) maps such that the homotopy classes $[f_\alpha]$ generate $\pi_1(X)$. We use these maps to glue 2-cells e_α^2 to X by making the amalgamated sum

$$X_1 = X \cup \bigcup_{\alpha \in A} e_\alpha^2 \ .$$

The space $\pi_1(X_1)$ is obviously simply-connected (Van Kampen theorem) and therefore the Hurewicz map $\rho : \pi_2(X_1) \longrightarrow H_2(X_1)$ is an isomorphism.

Consider the following commutative diagram associated to the inclusion $X \hookrightarrow X_1$

$$
\begin{array}{ccccc}
\pi_2(X_1) & \longrightarrow & \pi_2(X_1, X) & \longrightarrow & \pi_1(X) \\
\Big\downarrow{\scriptstyle p}{\simeq} & & \Big\downarrow & & \Big\downarrow \\
H_2(X_1) & \xrightarrow{\ j\ } & H_2(X_1, X) & \longrightarrow & H_1(X) = 0
\end{array}
$$

We have $H_1(X) = 0$ because $H_1(X)$ is $\pi_1(X)_{ab}$ and $\pi_1(X)$ is perfect by hypothesis. Therefore j is a surjective map.

By construction $H_2(X_1, X)$ is a free abelian group generated by (e_2^α), $\alpha \in A$. Choose $b_\alpha : S^2 \longrightarrow X_1$ such that $j \circ \rho[b_\alpha] = (e_2^\alpha)$. This is possible because j is surjective and ρ an isomorphism.

Using the maps b_α we glue 3-cells e_α^3 on X_1 to get the amalgamated sum

$$X^+ = X_1 \cup \bigcup_{\alpha \in A} e_\alpha^3 \ .$$

It is immediate to see that $\pi_1(X^+)$ is trivial (Van Kampen theorem). To show that the inclusion $i : X \longrightarrow X^+$ is an isomorphism in homology it suffices to prove that the relative homology $H_*(X^+, X; \mathbb{Z})$ is trivial.

The chain complex $C_*(X^+, X)$ which computes this homology has the following form:

$$\ldots \longrightarrow 0 \longrightarrow C_3(X^+, X) \xrightarrow{\ d\ } C_2(X^+, X) \longrightarrow 0 \longrightarrow \ldots$$

where $C_3(X^+, X)$ (resp. $C_2(X^+, X)$) is free abelian generated by the (e_α^3)'s (resp. (e_α^2)'s), $\alpha \in A$. Therefore it is sufficient to prove that $d(e_\alpha^3) = (e_\alpha^2)$. But by construction the boundary of (e_α^3) is the image of the map b_α, and this image is

$$j \circ \rho[b_\alpha] = (e_\alpha^2) \in H_2(X_1, X) = C_2(X_1, X) = C_2(X^+, X) \ .$$

So the complex $C_*(X^+, X)$ is acyclic and we are done.   □

Here is a slight generalization of this theorem.

Theorem 5.2. *Let X be a connected cellular complex and N a normal perfect subgroup of $\pi_1(X)$. Then there exists a space X^+ (depending on N) and a map $i : X \longrightarrow X^+$ such that*

a) $i_* : \pi_1(X) \longrightarrow \pi_1(X^+)$ is the surjection $\pi_1(X) \longrightarrow \pi_1(X)/N$,
b) for any $\pi_1(X)/N$-module L there is an isomorphism $i_* : H_*(X, i^*(L)) \longrightarrow H_*(X^+, L)$ where $i^*(L)$ is L considered as a $\pi_1(X)$-module.

Proof. Let $Y \longrightarrow X$ be the covering associated to the surjection $\pi_1(X) \longrightarrow \pi_1(X)/N$. Therefore $\pi_1(Y) = N$ and we can apply the plus-construction to Y. The space X^+ is defined to be the amalgamated sum in the following diagram

$$\begin{array}{ccc} Y & \longrightarrow & Y^+ \\ \downarrow & & \downarrow \ . \\ X & \longrightarrow & X^+ \end{array}$$

The first property follows from the Van Kampen theorem and 5.1, the second one follows from excision property for a push-out diagram.   □

Proposition 5.3. *Let X be a connected space and N a normal perfect subgroup of $\pi_1(X)$. Let $f : X \longrightarrow Y$ be a continuous map such that $f_*(N) = 1$. Then there exists a unique factorization f up to homotopy*

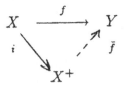

Proof. This is an immediate application of obstruction theory.   □

Example 1. Let X be a space which is a homology n-sphere, i.e.

$$H_*(X, \mathbb{Z}) = \begin{cases} \mathbb{Z} & \text{if } * = 0 \\ 0 & \text{if } * \neq 0, n \quad \text{for } n > 1 \ . \\ \mathbb{Z} & \text{if } * = n \end{cases}$$

As $H_1(X, \mathbf{Z}) = \pi_1(X)_{ab}$ is trivial, $\pi_1(X)$ is perfect. The space X^+ is simply connected and $H_*(X^+) = H_*(S^n)$, therefore by Whitehead's theorem X^+ is homotopy equivalent to a sphere.

Example 2. Many interesting examples come from the following situation. Let G be a discrete group and BG its classifying space (in other words BG is an Eilenberg-Mac Lane space of type $K(G, 1)$). Suppose that $[G, G]$ is perfect (G is then called quasi-perfect) and apply the plus-construction to BG with $N = [G, G]$. Then the homotopy of BG^+ is very interesting (but also very difficult to compute). The main example $G = GL(A)$ will be treated later. Let us mention the case $G = \Sigma_\infty$ where $\Sigma_\infty = \lim_n \Sigma_n$, $\Sigma_n = $ symmetric group on n letters.

Then it was proved by Quillen that $B\Sigma_\infty^+$ is homotopy equivalent to $\lim_n \Omega^n S^n = (\Omega S^0)_0$.

In order to define algebraic K-groups we need some information on the general linear group $GL(A)$.

The group of invertible $n \times n$-matrices with entries in the ring A is denoted $GL_n(A)$. It is viewed as the subgroup of matrices of $GL_{n+1}(A)$ of the form

$$\begin{pmatrix} & & & 0 \\ & M & & \vdots \\ & & & 0 \\ 0 & \cdots & 0 & 1 \end{pmatrix}$$

The union $\lim_n GL_n(A)$ is the infinite general linear group $GL(A)$.

Let e_{ij}^a be the matrix with 1's on the diagonal, a at the (i, j)-place and 0 otherwise ($i \neq j$). The normal subgroup of $GL_n(A)$ (resp. $GL(A)$) generated by these matrices is called the *elementary group* and denoted $E_n(A)$ (resp. $E(A)$).

Lemma 5.4. Whitehead's Lemma [M]. *The groups $E_n(A)$ ($n \geq 3$) and $E(A)$ are perfect and $E(A)$ is the commutator subgroup $[GL(A), GL(A)]$.*

Proof. It is immediate that if i, j, k are three different indices, then one has $\left[e_{ij}^a, e_{jk}^b\right] = e_{ik}^{ab}$. Therefore $E_n(A)$ is perfect as soon as $n \geq 3$. As a consequence $E(A) \subset [GL(A), GL(A)]$. We now prove the converse inclusion. If α is an invertible matrix then

$$w(\alpha) \equiv \begin{bmatrix} 0 & \alpha \\ -\alpha^{-1} & 0 \end{bmatrix} = \begin{bmatrix} 1 & \alpha \\ 0 & 1 \end{bmatrix} \begin{bmatrix} 1 & 0 \\ -\alpha^{-1} & 1 \end{bmatrix} \begin{bmatrix} 1 & \alpha \\ 0 & 1 \end{bmatrix}$$

and

$$w(\alpha)w(-1) = \begin{bmatrix} \alpha & 0 \\ 0 & \alpha^{-1} \end{bmatrix} .$$

This proves that the diagonal matrix $\begin{bmatrix} \alpha & 0 \\ 0 & \alpha^{-1} \end{bmatrix}$ is in $E(A)$. Let α and β be in $GL_n(A)$, then in $GL_{3n}(A)$ we have the following identity

$$\begin{bmatrix} \alpha\beta\alpha^{-1}\beta^{-1} & & \\ & 1 & \\ & & 1 \end{bmatrix} = \begin{bmatrix} \begin{bmatrix} \alpha & & \\ & \alpha^{-1} & \\ & & 1 \end{bmatrix}, \begin{bmatrix} \beta & & \\ & 1 & \\ & & \beta^{-1} \end{bmatrix} \end{bmatrix} .$$

This proves that $[GL(A), GL(A)] \subset E(A)$.

Thanks to Whitehead's lemma we can apply Quillen's plus-construction to the classifying space $BGL(A)$ with respect to the normal perfect subgroup $E(A)$ to get $BGL(A)^+$.

Definition 5.5. For $n \geq 1$ the *algebraic K-groups of* A are

$$K_n(A) = \pi_n\left(BGL(A)^+\right) .$$

Remark. In order to make this construction functorial we first make a choice for $A = \mathbb{Z}$ and then define $BGL(A)^+$ as the amalgamated sum

$$BGL(A)^+ = BGL(A) \cup_{BGL(\mathbb{Z})} BGL(\mathbb{Z})^+ .$$

If we only need to compute $K_n(A)$ rationally then there is a homological interpretation of $K_n(A) \otimes \mathbb{Q}$ which is quite helpful for computations.

The direct sum of matrices defines a map $GL(A) \times GL(A) \longrightarrow GL(A)$ and therefore a map $BGL(A)^+ \times BGL(A)^+ \longrightarrow BGL(A)^+$. Hence it can be shown that $BGL(A)^+$ inherits the structure of an H-space. It is well-known that the rational homotopy of an H-space is the primitive part of its homology. But $H_*\left(BGL(A)^+, \mathbb{Q}\right) = H_*\left(BGL(A), \mathbb{Q}\right) = H_*\left(GL(A), \mathbb{Q}\right)$. Therefore we have the following.

Theorem 5.6. $K_n(A) \otimes \mathbb{Q} = \operatorname{Prim} H_n\left(GL(A), \mathbb{Q}\right)$.

$\square$

6. K_1, K_2 and K_3

6.1. $K_1(A)$

By properties of the plus-construction we have

$$K_1(A) = GL(A)_{ab} = GL(A)/E(A) .$$

In terms of homology of discrete groups we have $K_1(A) = H_1\left(GL(A), \mathbb{Z}\right)$. If A is commutative the determinant factors through $GL(A)_{ab}$, therefore

$$\det : GL(A) \longrightarrow K_1(A) \longrightarrow A^* .$$

As $A^* = GL_1(A)$, it splits in $K_1(A)$ and the remaining part is denoted $SK_1(A) : K_1(A) = A^* \times SK_1(A)$.

For many rings (fields, rings of integers in a number field, local rings) $SK_1(A)$ is trivial. However this is not always the case. For instance if $A = \mathbb{Z}[x, y, z, t]/(xt - yz = 1)$ then the matrix $\begin{bmatrix} x & y \\ z & t \end{bmatrix}$ is an element of $GL_2(A)$ whose determinant is trivial and which generates an infinite cyclic element in $SK_1(A)$ (in fact $SK_1(A) \simeq \mathbb{Z}$ here).

6.2. $K_2(A)$

By definition of $BGL(A)^+$ we know that its universal cover is $BE(A)^+$. Therefore

$$K_2(A) = \pi_2 BGL(A)^+ = \pi_2(BE(A)^+)$$
$$= H_2(BE(A)^+) = H_2(BE(A)) = H_2(E(A), \mathbb{Z}) \ .$$

In this sequence of equalities we use Hurewicz theorem and property b) of the plus-construction.

There is another way to look at $K_2(A)$. Consider the homotopy fibration

$$F_A \longrightarrow BE(A) \longrightarrow BE(A)^+ \ .$$

It gives rise to an exact sequence of groups

$$1 \longrightarrow K_2(A) \longrightarrow \pi_1 F_A \longrightarrow E(A) \longrightarrow 1 \ .$$

From the universal property of the plus-construction we can prove that this extension is central and universal among the central extensions of $E(A)$.

On the other hand Steinberg has constructed a group $St(A)$ by generators and relations as follows:

$$\text{generators:} \quad x_{ij}^a \quad i \neq j, a \in A$$
$$\text{relations: 1.} \quad x_{ij}^a \, x_{ij}^b = x_{ij}^{a+b}$$
$$2. \quad [x_{ij}^a, x_{kl}^b] = 1 \quad i \neq l, h \neq k$$
$$3. \quad [x_{ij}^a, x_{jk}^b] = x_{ik}^{ab} \quad i \neq k$$

The map $\phi : St(A) \longrightarrow E(A)$, $x_{ij}^a \longmapsto e_{ij}^a$ is a surjective group homomorphism. In fact it is the universal central extension of $E(A)$ (cf. [M]). Therefore $\pi_1(F(A)) = St(A)$ and

$$K_2(A) = \ker(\phi : St(A) \longrightarrow E(A)) \ .$$

This identification is one of the main tools to compute K_2-groups.

We now look at the particular case where A is a field F. Let $u \in F^*$ and put $w_{12}(u) = x_{12}^u \, x_{21}^{-u^{-1}} \, x_{12}^u$. Then the image of $h_{12}(u) = w_{12}(u)w_{12}(-1)$ in $E(A)$ is the diagonal matrix $\begin{bmatrix} u & 0 \\ 0 & u^{-1} \end{bmatrix}$. Similarly we define $h_{13}(v)$ for $v \in F^*$. The commutator $[h_{12}(u), h_{13}(v)]$ maps to 1 in $E(A)$ since diagonal matrices commute, hence this element is in $K_2(A)$. One can show that it does not depend on the indices 1, 2 and 3 so we can denote it $\{u,v\}$. Computations in $St(A)$ prove that the following relations are satisfied:

a. $\{uu',v\} = \{u,v\}\{u',v\}$, $\quad \{u,vv'\} = \{u,v\}\{u,v'\}$
b. $\{u,1-u\} = 1$, $\quad u \neq 1$.

Definition . The Milnor K_2 is the group $K_2^M(A)$ defined by generators $\{u,v\}$, $u,v \in F^*$ and the above relations.

Theorem 6.3. Matsumoto's theorem [M]. *The map $K_2^M(F) \longrightarrow K_2(F)$ is an isomorphism.*

We refer to [M] for the proof (still valid for $A = \mathbb{Z}$). It has been generalized by Maazen and Stienstra for local rings under the following form. Let

$$\langle a,b \rangle = x_{21}^{-b(1-ab)^{-1}} x_{12}^{-1} x_{21}^{b} x_{12}^{(1-ab)^{-1}a} h_{12}(1-ab)^{-1}$$

where a and b in A are such that $1-ab$ is invertible. Computations in $St(A)$ prove that the following relations hold (written additively, i.e. in $K_2(A)$)

a) $\langle 1,b \rangle = 1$
b) $\langle a,b \rangle + \langle a',b \rangle = \langle a+a'-aba',b \rangle$
c) $\langle ab,c \rangle - \langle a,bc \rangle + \langle ca,b \rangle = 0$.

Definition . $K_2^M(A)$ is the abelian group with generators $\langle a,b \rangle$ for $a,b \in A$ such that $1-ab \in A^*$ and with the three relations above.

When $A = F$ the two definitions of $K_2^M(F)$ agree thanks to $\{u,v\} \longmapsto \langle u,(1-v)u^{-1} \rangle$.

Theorem 6.4. (Maazen-Stienstra [M-S]). *If A is a local ring $K_2^M(A) \longrightarrow K_2(A)$ is an isomorphism.*

Remark. In relation b) we note that $(1-ab)\cdot(1-a'b) = (1-(a+a'-aba')b)$. So this relation comes in fact from multiplication in A. If instead we take addition, that is we replace b) by b') $\langle a,b \rangle + \langle a',b \rangle = \langle a+a',b \rangle$, then the group that we obtain is $K_2^+(A) = HC_1(A)$.

6.5. $K_3(A)$

We now go back to the exact sequence

$$1 \longrightarrow K_2(A) \longrightarrow St(A) \longrightarrow E(A) \longrightarrow 1 .$$

Taking classifying spaces it gives rise to a fibration

$$BK_2(A) \longrightarrow BSt(A) \longrightarrow BE(A) .$$

Proposition 6.6. *Applying the plus-construction still gives a fibration*

$$BK_2(A) \longrightarrow BSt(A)^+ \longrightarrow BE(A)^+ .$$

Sketch of proof: It is a comparison of spectral sequences, plus the properties of the plus-construction. □

As a consequence we have

$$K_3(A) = \pi_3\big(BE(A)^+\big) = \pi_3\big(BSt(A)^+\big)$$
$$= H_3\big(BSt(A)^+\big) = H_3\big(BSt(A)\big) = H_3\big(St(A), \mathbb{Z}\big)$$

Summarizing these homological computations gives:

$$K_1(A) = H_1\big(GL(A), \mathbb{Z}\big)$$
$$K_2(A) = H_2\big(E(A), \mathbb{Z}\big)$$
$$K_3(A) = H_3\big(St(A), \mathbb{Z}\big) .$$

There is no natural candidate for K_4 as far as I know.

7. Product and Milnor K-Theory

Let us take A to be a commutative ring. We define a product on K-theory from the tensor product of matrices:

$$\otimes : GL_n(A) \times GL_p(A) \longrightarrow GL_{np}(A)$$

(Remark that this necessitates the choice of an isomorphism $A^{\otimes n} \otimes A^{\otimes p} \xrightarrow{\sim} A^{\otimes np}$). This map is *not* compatible with stabilization and therefore does not extend to the infinite general linear group. However it defines a map:

$$\gamma_{n,p} : BGL_n(A)^+ \times BGL_p(A)^+ \longrightarrow BGL_{np}(A)^+ \longrightarrow BGL(A)^+ .$$

As $BGL(A)^+$ is an H-space we can form a new map

$$\gamma_{n,p} = \hat{\gamma}_{n,p}(x, y) = \gamma_{n,p}(x, y) - \gamma_{n,p}(x, *) - \gamma_{n,p}(*, y) .$$

And now one can check that this new map commutes with stabilization up to homotopy (cf. [L1]) and gives

$$\hat{\gamma} : BGL(A)^+ \times BGL(A)^+ \longrightarrow BGL(A)^+ .$$

Theorem 7.1. *[L1]. The map $\hat{\gamma}$ induces a product $* : K_n(A) \times K_m(A) \longrightarrow K_{n+m}(A)$ which is graded commutative.*

Consider now the tensor product $F^* \otimes_{\mathbb{Z}} \ldots \otimes_{\mathbb{Z}} F^*$ where F is a field. Milnor K-theory is defined as follows:

Definition 7.2. $K_n^M(F) = F^* \otimes \ldots \otimes F^* / \{(a_1 \otimes \ldots \otimes a_n) \mid a_i + a_j = 1 \text{ for some } i \neq j\}$.

It is immediate that $K_1^M(F) = F^*$ and $K_2^M(F)$ is the same group as the one considered above. The product structure permits us to define a map

$$i : K_n^M(F) \longrightarrow K_n(F)$$

by sending the class of $(a_1 \otimes \ldots \otimes a_n)$ to $a_1 * \ldots * a_n$.

As noted before this is an isomorphism for $n = 1, 2$.

Theorem 7.3. (Suslin [S]). *There is a map $p : K_n(F) \longrightarrow K_n^M(F)$ such that $p \circ i$ is multiplication by $(n-1)!$.*

This is in fact a consequence of more subtle result about stability of the homology of the general linear group.

Theorem 7.4. (Suslin [S]). *Let F be an infinite field, then the following maps are isomorphisms*

$$H_n(GL_n(F)) \longrightarrow H_n(GL_{n+1}(F)) \longrightarrow \ldots \longrightarrow H_n(GL(F))$$

and there is an exact sequence

$$H_n(GL_{n-1}(F)) \longrightarrow H_n(GL_n(F)) \longrightarrow K_n^M(F) \longrightarrow 0 \ .$$

The map p is obtained from this theorem as follows:

$$K_n(F) \overset{\text{Hurewicz}}{\longrightarrow} H_n(GL(F)) \overset{\sim}{\longleftarrow} H_n(GL_n(F)) \longrightarrow K_n^M(F) \ .$$

8. K-Theory of Nilpotent Ideals

Recently T. Goodwillie [G] announced a very interesting theorem about K-theory of nilpotent ideals.

Theorem 8.1. *Let A be a commutative ring and I a nilpotent ideal, then there is an isomorphism*

$$K_n(A, I) \otimes \mathbb{Q} \longrightarrow HC_{n-1}(A, I) \otimes \mathbb{Q} \ .$$

Here the relative K-groups $K_n(A, I)$ are defined as $\pi_n(F_{A,I})$ where $F_{A,I}$ is the homotopy fiber of $BGL(A)^+ \longrightarrow BGL(A/I)^+$.

This theorem is immediate for $n = 1$ because $K_1(A, I) = (1 + I)^*$ and $HC_0(A, I) = I$. Therefore the isomorphism is given by

$$(1-x) \longmapsto L(x) = -\sum_{n>0} \frac{x^n}{n} \ .$$

For $n = 2$ $K_2(A, I)$ admits a presentation using the Dennis-Stein symbols $\langle -, - \rangle$ as follows: Generators are $\langle a, b \rangle$ with $a, b \in A$ and a or $b \in I$ and with the relations a), b) and c) of chapter 6.

The group $HC_1(A, I)$ has a similar presentation but with relation b) replaced by relation b').

The isomorphism $K_2(A, I) \otimes \mathbb{Q} \longrightarrow HC_1(A, I) \otimes \mathbb{Q}$ is proved by using the map

$$\langle a, b \rangle \longmapsto \langle \frac{L(ab)}{b}, b \rangle \quad \text{for} \quad a \in I, b \in A \ .$$

Relation c) is easy to check. Relation b) is a consequence of the functional equation of the logarithm:

$$L(x) + L(y) = L(x + y - xy) \ .$$

Relation a) is proved as follows. We need to show that $\langle \frac{L(u)}{u}, u \rangle = 0$ in $HC_1(A, I)$ when $u \in I$. Using c) and b') we see that $\langle 1, u^n \rangle = n \langle u^{n-1}, u \rangle$. Therefore

$$\langle \frac{L(u)}{u}, u \rangle = \sum \frac{1}{n} \langle u^{n-1}, u \rangle = \sum \frac{1}{n^2} \langle 1, u^n \rangle = \langle 1, L_2(u) \rangle$$

where $L_2(u)$ is the dilogarithm function $\sum \frac{u^n}{n^2}$ (remark that all these sums are finite because u is nilpotent). Then by a) we get $\langle 1, L_2(u) \rangle = 0$.

The isomorphism in the other way is given by $\langle \frac{1 - \exp(xy)}{y}, y \rangle \longmapsto \langle x, y \rangle$ for $y \in A$ and $x \in I$.

Another explicit case is when $A = R[\varepsilon]/\varepsilon^2 = 0$ (dual numbers) and $I = (\varepsilon)$. Then $HC_n(A, I)$ contains $\Omega^{n-1}_{R/\mathbb{Z}}$ as a direct factor and so does $K_n(A, I)$ as proved in [L3].

Let A be a truncated polynomial ring $A = k[t]/t^{\ell+1}$. Then, Goodwillie's theorem, plus the computation of cyclic homology of A (cf. prop. 4.2) gives a complete answer for the computation of $K_n(A)/K_n(k)$.

In the case $\ell = 1$ and $k = \mathbb{Z}$ this computation was done by Soulé and I constructed explicitly a generator in $K_{2n+1}(\mathbb{Z}[\varepsilon])$ (cf. [L3]). A similar, but more complicated, construction can be performed for the generators of $K_n\left(k[t]/t^{\ell+1}\right)$.

The computation of $K_n\left(\mathbb{Z}[t]/t^{\ell+1}\right)$ is much more difficult and complete results are known in low dimensions (cf. [A]).

References

[A] Aisbett, J.E., Lluis-Puebla, E., Snaith, V.: On $K_*(\mathbb{Z}/n)$ and $K_*\left(F_q[t]/(t^2)\right)$. Memoirs AMS vol. 57, N° 329, 1985

[C] Connes, A.: Non commutative differential geometry. Publ. Math. IHES **62** (1985) 257–360

[D-P] de Concini, C., Procesi, C.: A characteristic free approach to invariant theory. Advances in Math. **21** (1976) 330–354

[F-T] Feigin, B.L., Tsygan, B.L.: Additive K-theory. Springer Lecture Notes 1289 (1987) 67–209

[G]   Goodwillie, T.G.: Cyclic homology, derivations and the free loop space. Topology **24** (1985) 187–215

[K-L] Kassel, C., Loday, J.-L.: Extensions centrales d'algèbres de Lie. Ann. Inst. Fourier **33** (1982) 119–142

[L1] Loday, J.-L.: K-théorie algèbrique et représentations de groupes. Ann. Sc. ENS **9** (1976) 309–377

[L2] Loday, J.-L.: Cyclic homology: a survey, in geometric and algebraic topology. Banach Center Publications, vol. 18, Warsawa (1986) 285–307

[L3] Loday, J.-L.: Symboles en K-théorie algèbrique supérieure. CRAS **292** (1981) 863–866

[L-Q] Loday, J.-L., Quillen, D.: Cyclic homology and the Lie algebra homology of matrices. Comm. Math. Helv. **59** (1984) 565–591

[M-S] Maazen, H., Stienstra, H.: A presentation of split radical pairs. J. Pure Applied Algebra **10** (1977/78) 271–294

[M]   Milnor, J.: Introduction to algebraic K-theory. Annals of Mathematics Studies, N° 72. Princeton University Press, 1971

[Q]   Quillen, D.: Cohomology of groups. Actes du Congrès International des Mathématiciens (Nice, 1970), t. 2, pp. 47–51

[S]   Suslin, A.A.: Homology of GL_n, characteristic classes and Milnor K-theory. In: Alg. K-theory, number theory, geometry and analysis. Springer Lecture Notes in Math., 1046 (1984) 357–375

[T]   Tsygan, B.L.: The homology of matrix Lie algebras over rings and the Hochschild homology. Uspekhi Mat. Nauk bf 38 (1983) 217–218

[W]   Weyl, H.: The classical groups. Princeton University Press, 1946

Comparing Algebraic and Topological K-Theory

Henri Gillet

Department of Mathematics, University of Illinois Chicago, Box 4348, Chicago, Ill. 60680, USA

The interplay between algebraic and topological K-theory has played an important role in the development of both fields. Two important conjectures in algebraic geometry, the Hodge and Tate conjectures, can be phrased as assertions about the image of the natural transformation from algebraic to topological K-theory, while the conjectures of Lichtenbaum and Quillen on the K-theory of rings of algebraic integers and on the K-theory of algebraically closed fields were motivated in part by analogies with topological K-theory. In these notes I have tried to describe some aspects of the relationship, finishing with a proof of Suslin's rigidity theorem.

The rigidity theorem is half of Suslin's proof of Quillen's conjecture on the K-theory of algebraically closed fields, the other half of which is his computation of the torsion in the K-theory of the complex numbers, which is described in Soulé's lectures. In general, rather than giving detailed proofs, I have tried to give comprehensive references. I hope that the reader will be encouraged to look at the original sources.

In section 1 I give a brief review of K_0 of a ring, the definition of an exact category and its K_0. Section 2 is a discussion of topological K-theory, starting with a description of the category of vector bundles on a topological space. The section continues with the theorem of Serre and Swan on the equivalence of the category of vector bundles on a compact space X and the category of projective modules over the ring $C(X)$ of continuous functions on X. The section finishes with a discussion of the K-theory of compact spaces. Section three contains the definition of the functor K_0 on the category of schemes. It should be observed that we have reversed the historical order here. The functor K_0 was first defined for algebraic varieties, then for topological spaces. Higher algebraic K-theory of varieties and schemes is introduced in section four, using Quillen's Q-construction applied to the category of vector bundles. Section five describes the natural transformation from algebraic K_0 to topological K^0, while section six does the same for higher K-theory. In both §6 and §7 we discuss the information that Hodge theory gives us about the image of algebraic K-theory in topological K-theory: for K_0 the image is (conjecturally) described by the Hodge conjecture, while for higher K-theory, the image must be torsion. This leads us in section 7 to consider K-theory with coefficients finishing with a proof of Suslin's rigidity theorem.

Finally I would like to thank E. Lluis-Puebla and the Universidad Nacional Autónoma de México for organizing the orignal lecture series, and for their patience during the delay in the preparation of these notes.

1. What is Algebraic K-Theory?

1.1

A basic theorem of linear algebra is:

Theorem . *Let k be a field, and $A = (a_{ij})$ an $m \times n$ matrix with entries in k. Then the solutions of the system of equations*

$$a_{11}x_1 + \quad \ldots \quad + a_{1n}x_n = 0$$

$$\vdots$$

$$a_{m1}x_1 + \quad \ldots \quad + a_{mn}x_n = 0$$

i.e.

$$A\underline{x} = \underline{0}$$

form a vector space S over k, and if R is the subspace of k^m (= the space of column vectors of length m) spanned by columns of A, then

$$\dim S + \dim R = n \ .$$

Another way of stating this result is:

1.2

Theorem . *Let k be a field.*

i) Every finitely generated k module is free, with well defined dimension or rank.

ii) If

$$0 \longrightarrow U \overset{\alpha}{\longrightarrow} V \overset{\beta}{\longrightarrow} W \longrightarrow 0$$

is a short exact sequence of vector spaces, i.e. α and β are linear transformations with $Kernel\,(\alpha) = 0$, $Image\,(\alpha) = kernel\,(\beta)$, and $W = Image\,(\beta)$, then

$$\dim U + \dim W = \dim V \ .$$

Algebraic K-Theory is in part an attempt to understand how this theorem may be generalized to arbitrary rings. For simplicity I will consider only commutative rings (with units) in what follows.

1.3

Recall that if R is a ring, an R-module P is *projective* if the following equivalent conditions are satisfied:

P1) There exists a module Q such that $P \oplus Q$ is free.
P2) For any surjective homomorphism of modules $f : M \longrightarrow N$, given any homomorphism $\phi : P \longrightarrow N$, there exists a homomorphism $\tilde{\phi} : P \longrightarrow M$ such that $f \cdot \tilde{\phi} = \phi$.

If k is a field, then every finitely generated k-module is projective, indeed free. If k is a Dedekind domain, such as the ring of integers in a number field, then every finitely generated torsion free module is projective, but will not in general be free.

1.4

Observe that by P1 above, if P_1 and P_2 are projective modules over the ring R, then so is $P_1 \oplus P_2$. Let $\mathcal{P}$ equal the set of isomorphism classes of finitely generated projective R-modules.

The operation $\oplus$ makes $\mathcal{P}$ a commutative monoid i.e. a set with an associative, commutative law of addition possessing a zero element, but in which elements may not have additive inverses; i.e. if we write $\{P\}$ for the isomorphism class of a projective R-module, then we set

$$\{P_1\} + \{P_2\} = \{P_1 \oplus P_2\} \ .$$

Note that by P2, if

$$0 \longrightarrow M \xrightarrow{\alpha} N \xrightarrow{\beta} P \longrightarrow 0$$

is a short exact sequence of projective R-modules, then there is a splitting $\eta : P \longrightarrow N$ such that $\beta \cdot \eta = I_P$, and an induced isomorphism

$$\alpha \oplus \eta : M \oplus P \xrightarrow{\sim} N \ .$$

Hence we can also say that in P

$$\{M\} + \{P\} = \{N\}$$

whenever there is a short exact sequence as above.

Now in general, given a commutative monoid M, there is an associated group $\hat{M}$, the *group completion* of M, defined by a process analogous to the formation of the integers from the natural numbers, or of the localization of a ring with respect to a multiplicative subset.

Proposition . *If M is a commutative monoid, let $\hat{M}$ be the set with elements represented by pairs (m_1, m_2) with $m_i \in M$, and $(m_1, m_2) = (n_1, n_2)$ if there exists $p \in M$ with $m_1 + n_2 + p = n_1 + m_2 + p$. (Note that in the absence of a cancellation law in M, this last equation does not imply that $m_1 + n_2 = n_1 + m_2$). Then $\hat{M}$ is a group, and the map $\eta : M \longrightarrow \hat{M}$, $\eta(m) = (m, 0)$, has the property* *that if $f : M \longrightarrow G$ is any homomorphism of commutative monoids with G a group, f factors as $f = \bar{f} \cdot \eta$, where $\bar{f} : \hat{M} \longrightarrow G$ is a group homomorphism given by*

$$\bar{f}(m_1, m_2) = f(m_1) - f(m_2) \ .$$

Proof. Exercise. □

Definition . If R is a ring, $K_0(R) = \hat{\mathcal{P}}$, so $K_0(R)$ has elements $(\{P\}, \{Q\}) = [P] - [Q]$, where $[P] = (\{P\}, 0)$.

It is easy to see that the above definition is equivalent to the following: $K_0(R)$ is the quotient of the free abelian group on $\mathcal{P}$ by the subgroup generated by elements of the form $\{M\} + \{P\} - \{N\}$ for each exact sequence $0 \longrightarrow M \longrightarrow P \longrightarrow N \longrightarrow 0$ of projective modules.

Example . If k is a field, or more generally a local ring, every projective k-module is free, and $K_0(k) = \mathbb{Z}$, via $[P] \longrightarrow rk(P)$.

1.5

If $f : R \longrightarrow S$ is a homomorphism of rings and P is a finitely generated projective R-module then $S \otimes_R P$ is a f.g. projective S-module. (This is an easy exercise using P1) and the fact that $S\otimes_R$ preserves direct sums.)

Furthermore if

$$0 \longrightarrow P_1 \longrightarrow P \longrightarrow P_2 \longrightarrow 0$$

is an exact sequence of f.g. projective R-modules, then $P \simeq P_1 \oplus P_2$, and

$$0 \longrightarrow S \otimes_R P_1 \longrightarrow S \otimes_R P \longrightarrow S \otimes_R P_2 \longrightarrow 0$$

is also exact. Hence there is an induced map

$$K_0(f) : K_0(R) \longrightarrow K_0(S)$$
$$[P] \longmapsto [S \otimes_R P] \ .$$

This makes K_0 a *covariant functor* from the category of commutative rings to the category of abelian groups. This is one of K-theory's most important properties.

Another important property of K_0 is that, for a *commutative* ring R, $K_0(R)$ is itself a ring. The product is given by:

$$[P] \cdot [Q] = [P \otimes_R Q] \ .$$

(Note that $P \otimes_R Q$ would not in general be an R-module unless R is commutative.) It is straightforward to check that if $F : R \longrightarrow S$ is a homomorphism of commutative rings, then $K_0(f)$ is itself a ring homomorphism. The key point is that there is an isomorphism of S-modules, for each pair P and Q of projective R-modules:

$$S \otimes_R (P \otimes_R Q) = (S \otimes_R P) \otimes_S (S \otimes_R Q) \ .$$

1.6

The reader may have noticed that the definition of K_0 did not use very many of the properties of projective modules as such. A more general setting for the definition is the notion of an *exact category* ([Q2] §2). That is, an additive category E, equipped with a class $\mathcal{E}$ of sequences

$$(*)\qquad\qquad 0 \longrightarrow M' \overset{i}{\longrightarrow} M \overset{j}{\longrightarrow} M'' \longrightarrow 0$$

with the following properties:

a) Any sequence in $\mathcal{M}$ isomorphic to a sequence in $\mathcal{E}$ is also in $\mathcal{E}$. For any M', M'' in $\mathcal{M}$ the sequence

$$0 \longrightarrow M' \overset{(Id,0)}{\longrightarrow} M' \oplus M'' \overset{pr_2}{\longrightarrow} M'' \longrightarrow 0$$

is in $\mathcal{E}$. For any sequence $(*)$ in $\mathcal{E}$, i is a kernel for j, and j is a cokernel for i. We call i an *admissible monomorphism*, and j an *admissible epimorphism*.

b) The class of admissible epimorphisms is closed under base-change by arbitrary maps in $\mathcal{M}$, while the class of admissible monomorphisms is closed under cobase-change by arbitrary maps.

c) If $f : M \longrightarrow M''$ has a kernel in $\mathcal{M}$, then for any map $g : N \longrightarrow M$ such that $f \cdot g$ is an admissible epimorphism, then f is an admissible epimorphism. Dually for admissible monomorphisms.

For such an E we define $K_0(E)$ to be the free abelian group on the objects of E, modulo the relation that if M, N, P are objects of E, then $[M]+[P] = [N]$ if there is an admissible exact sequence $M \rightarrowtail N \twoheadrightarrow P$. If $F : E_1 \longrightarrow E_2$ is an exact functor, i.e. an additive functor which maps admissible exact sequences in E_1 to admissible exact sequences E_2, then $[P] \longrightarrow [F(P)]$ induces a homomorphism $K_0(F) : K_0(E_1) \longrightarrow K_0(E_2)$.

2. Topological K-Theory

2.1. Vector Bundles

We start our discussion of topological K-theory with some preliminaries on vector bundles. For simplicity we shall restrict our attention to complex vector bundles. Good general references for this material are [At][Ka1].

If X is a topological space, a *continuous family of complex vector spaces* $\pi : E \longrightarrow X$ over X consists of:

1) A continuous map of topological spaces $\pi : E \longrightarrow X$.
2) Continuous maps:
   $\mu : \mathbb{C} \times E \longrightarrow E$ ("scalar multiplication")
   $\sigma : E \times_X E \longrightarrow E$ ("addition").

This data satisfies the following axioms:

v(1) For all $\lambda \in \mathbb{C}$, $v \in E$, $\pi(\lambda v) = \pi(v)$.

v(2) For all $(v, w) \in E \times_X E$, $\pi(\sigma(v, w)) = \pi(v) = \pi(w)$.

v(3) If given $x \in X$, we write $E_x = \pi^{-1}(x)$, then the maps $\mu_x : \mathbb{C} \times E_x \longrightarrow E_x$ and $\sigma_x : E_x \times E_x \longrightarrow E_x$ make E_x a complex vector space.

E_x is called the *fibre of E over x*.

A homomorphism of continuous families of vector spaces over X, $f : (\pi : E \longrightarrow X) \longrightarrow (\pi' : F \longrightarrow X)$ is a continuous map $f : E \longrightarrow F$ such that $\pi' \cdot f = \pi$ i.e. f maps fibres to fibres, and such that for all $x \in X$, $f_x : E_x \longrightarrow E_{f(x)}$ is a linear transformation. With this definition, continuous families of vector spaces form a category. If $\pi : E \longrightarrow X$ and $\pi' : F \longrightarrow X$ are two families of complex vector spaces, we can form their direct sum, $E \oplus F$, which is just $E \times_X F$ with scalar multiplication and addition defined so that $(E \times_X F)_x = E_x \oplus F_x$ for all $x \in X$. More generally, if we have maps

$$0 \longrightarrow E' \xrightarrow{f} E \xrightarrow{g} E'' \longrightarrow 0$$

between continuous families of vector spaces, which over each point $x \in X$ give an exact sequence of fibres, we say that we have a (short) exact sequence of families of vector spaces. (Note that we usually write E in place of $\pi : E \longrightarrow X$, and 0 is the family $\{0\} \times X$.)

If $f : X \longrightarrow Y$ is a continuous map and $E = \pi : E \longrightarrow Y$ is a continuous family of vector spaces, we can form the pull-back $f^*(E) = E \times_Y X \longrightarrow X$. This is a continuous family of vector spaces over X. Clearly f^* is a functor from the category of continuous families of vector spaces over Y to the similar category for X. It is easy to see that f^* preserves exact sequences, and that if $g : Y \longrightarrow Z$ is another continuous map, then there is an isomorphism of functors $f^* g^* = (gf)^*$.

A family of vector spaces over X is *trivial* if it is isomorphic to a family $\mathbb{C}^n \times X \longrightarrow X$ (with the map equal to projection onto the second factor) for some $n \geq 0$. We say $\pi : E \longrightarrow X$ is a *vector bundle* if it is locally trivial, i.e., there is an open cover $X = \cup_i U_i$, and isomorphisms for each open set U_i:

$$\Theta_i : \mathbb{C}^{n_i} \times U_i \xrightarrow{\sim} E \mid_{U_i}$$

from a trivial bundle to the restriction of E to U_i.

It is straight forward to check that the direct sum of vector bundles is a vector bundle and that the pullback of a vector bundle is again a vector bundle.

If $\pi : E \longrightarrow X$ is a vector bundle, and $\Theta_i : \mathbb{C}^{n_i} \times U_i \longrightarrow E \mid_{U_i}$ family of local trivializations over the elements of an open cover $\{U_i\}$ of X, we have for each pair (i, j) an automorphism $\tilde{\Theta}_{ij} = \Theta_i^{-1} \cdot \Theta_j$ of $\mathbb{C}^{n_j} \times (U_i \cap U_j)$, i.e. a continuous map $\Theta_{ij} : U_i \cap U_j \longrightarrow GL_{n_j}(\mathbb{C})$. Note that $U_i \cap U_j \neq \emptyset$ implies that $n_i = n_j$. These maps satisfy the cocycle condition:

$$\Theta_{ij} \cdot \Theta_{jk} = \Theta_{ik}$$

for all triples (i, j, k). Conversely, one can show that if $\{U_i\}$ is an open cover of X, and $\Theta_{ij} : U_i \cap U_j \longrightarrow GL_{n_j}(\mathbb{C})$ a family of continuous maps satisfying the cocycle condition, then $\{\Theta_{ij}\}$ is the cocycle associated to a family of local trivializations of a vector bundle E over X. Explicitly:

$$E = E(\{\Theta_{ij}\}) = \left(\coprod_i \mathbb{C}^{n_i} \times U_i \right) / \sim$$

where $(v, x) \sim (\Theta_{ij}(v), x)$ if

$$(v, x) \in \mathbb{C}^{n_i} \times U_i \quad \text{and} \quad x \in U_i \cap U_j \ .$$

If E is a vector bundle over X, and $\{\Theta_{ij}\}$ is the cocycle associated to some local trivialization of E, it is straight forward to check that E is canonically isomorphic to the bundle $E(\{\Theta_{ij}\})$ determined by the cocycle.

Definition . i) Let X be a topological space. $\mathrm{Vect}(X)$ is the full subcategory of the category of continuous families of complex vector spaces over X consisting of the vector bundles and the morphisms between them. Note that $\mathrm{Vect}(X)$ is an exact category.

ii) $K^0(X) = K_0(Vect(X))$, i.e. $K^0(X)$ is the quotient of the free abelian group on the set of isomorphism classes of vector bundles over X by the group of relations generated by all expressions $([E] - [E'] - [E''])$ for $0 \longrightarrow E' \longrightarrow E \longrightarrow E'' \longrightarrow 0$ a short exact sequence of bundles. We shall also denote this group $K^0_{\mathrm{top}}(X)$ to distinguish it from the algebraic K-theory which shall appear later.

2.2. Sheaves of Sections

If $\pi : E \longrightarrow X$ is a continuous family of vector spaces over X, and $Z \subset X$ is a subset, we write $\Gamma(Z, E)$ for the set of continuous sections $f : Z \longrightarrow E$, i.e. continuous maps such that $\pi \cdot f : Z \longrightarrow X$ is the inclusion. If $\lambda : Z \longrightarrow \mathbb{C}$ is a continuous function define $\lambda \cdot f = \mu \cdot (\lambda, f)$, while if f and g are two continuous sections of E over Z then define $f + g = \sigma \cdot (f, g)$. It is easily checked that with these operations of scalar multiplication and addition, the set $\Gamma(Z, E)$ of continuous sections of E over Z becomes a module over the ring $\mathcal{C}(Z)$ of continuous complex valued functions on Z. Notice that if $Y \subset Z$ then restriction gives a homomorphism of abelian groups $\rho_{Y,Z} : \Gamma(Z, E) \longrightarrow \Gamma(Y, E)$ which is a homomorphism of $\mathcal{C}(Z)$-modules if one views $\Gamma(Y, E)$ as a $\mathcal{C}(Z)$-module via the ring homomorphism $\mathcal{C}(Z) \longrightarrow \mathcal{C}(Y)$ given by restriction of functions.

We are particularly interested in $\Gamma(U, E)$ when $U \subset X$ is an *open* set. Recall (see [Go], [Ha]) that a *pre-sheaf* $\mathcal{F}$ of sets (or groups, ... , etc.) on a topological space X consists of the following data:

i)   For each open subset $U \subset X$ a set (or group ...) $\mathcal{F}(U)$.

ii) To each inclusion of open sets $V \subset U$, a restriction function (or homomorphism ...)

$$\rho_{V,U} : \mathcal{F}(U) \longrightarrow \mathcal{F}(V)$$

such that, if $W \subset V \subset U$, $\rho_{W,V} \cdot \rho_{V,U} = \rho_{W,U}$ and for all U, $\rho_{U,U} = $ Identity.

If $U = \bigcup_{i \in I} U_i$ is a union of open subsets, write H for the subset of $\prod_{i \in I} \mathcal{F}(U_i)$ consisting of families $x_i \in \mathcal{F}(U_i)$ such that for all $i, j \in I \times I$ with $U_i \cap U_j \neq \emptyset$, then $\rho_{U_i \cap U_j, U_i}(x_i) = \rho_{U_i \cap U_j, U_j}(x_j) \in \mathcal{F}(U_i \cap U_j)$. Observe by axiom (ii) that there is a map $\rho : \mathcal{F}(U) \longrightarrow H$ given by $\rho(x) = \{x_i\}$ with $x_i = \rho_{U_i, U}(x) \in \mathcal{F}(U_i)$. We say $\mathcal{F}$ is a *sheaf* of sets (resp. groups, vector spaces, ...) if

iii) ρ is bijective (resp. an isomorphism).

Suppose that $\mathcal{R}$ is a sheaf of rings on X, and $\mathcal{M}$ a sheaf of abelian groups. We say that $\mathcal{M}$ is a sheaf of $\mathcal{R}$-modules if:

i) For all open subsets $U \subset X$, $\mathcal{M}(U)$ is an $\mathcal{R}(U)$- module
ii) for each inclusion $V \subset U$, the restriction map $\rho_{V,U} : \mathcal{M}(U) \longrightarrow \mathcal{M}(V)$ is $\mathcal{R}(U)$-linear, with respect to the $\mathcal{R}(U)$-module structure on $\mathcal{M}(V)$ induced by the homomorphism of rings $\rho_{V,U} : \mathcal{R}(U) \longrightarrow \mathcal{R}(V)$.

Note that on a topological space X, $U \longrightarrow \mathcal{C}(U)$ is a sheaf, the sheaf $\mathcal{C} = \mathcal{C}_X$ of continuous complex valued functions. The proof of the following lemma is left as an exercise:

Lemma . *Let $\pi : E \longrightarrow X$ be a continuous family of complex vector spaces over the topological space X. Then*

$$U \longrightarrow \Gamma(U, E)$$

is a sheaf of $\mathcal{C}_X$-modules.

If E is locally trivial, then the sheaf $U \longrightarrow \Gamma(U, E)$ is *locally free*, i.e. for each $x \in X$ there exists an open neighbourhood U of x such that $\Gamma(U, E)$ is a free $\mathcal{C}(U)$-module of finite rank.

Proposition . *The mapping $\Gamma : E \longrightarrow \Gamma(\_, E)$ is an exact fully faithful functor*

$$\Gamma : \mathrm{Vect}_X \longrightarrow \mathcal{M}\mathrm{od}_C$$

from the category of complex vector bundles to the category of locally free sheaves of $\mathcal{C}_X$-modules of finite rank.

2.3. Properties of the Category of Vector Bundles

Suppose that X is a paracompact space (i.e. it is Hausdorff and every open cover has a locally finite refinement). Then every open covering $\mathcal{U} = \{U_i\}_{i \in I}$ of X admits a *partition of unity* associated to $\mathcal{U}$, i.e.

There is a collection $\{\phi_i\}_{i \in I}$ of continuous real valued functions on X such that

i) $\phi_i(x) \geq 0$ for all $x \in X$.

ii) $x \notin U_i \Rightarrow \phi_i(x) = 0$.

iii) For each $x \in X$, there exists an open neighbourhood V of x such that all but finitely many of the ϕ_i vanish on V.

iv) For all $x \in X$, $\sum_{i \in I} \phi_i(x) = 1$. Note that this sum makes sense because of (iii).

Theorem . *i) Let X be a paracompact space; then any exact sequence of vector bundles*

$$0 \longrightarrow E' \overset{\alpha}{\longrightarrow} E \overset{\beta}{\longrightarrow} E'' \longrightarrow 0$$

splits, i.e., there exists a map $\sigma : E'' \longrightarrow E$ such that $\beta \cdot \sigma = \mathrm{Id}_{E''}$.

ii) If X is compact, and E is a vector bundle over X, then there exists a vector bundle E' over X such that $E \oplus E'$ is trivial, i.e. $E \oplus E' \simeq E_X^n$ for some n. (Here $E_X^n = \mathbb{C}^n \times X$.)

Proof. i) At each point $x \in X$, there is an open neighbourhood U_x in which $E \simeq E^r$ and $E'' \simeq E^s$, so β is represented by an $r \times s$ matrix with entries in $\mathcal{C}(U_x)$. At the point x there is an s-tuple of these columns which span the column space; i.e. if $\underline{e}_{i_1}, \ldots, \underline{e}_{i_s}$ are the corresponding sections of the trivial bundle E^r, $\beta(\underline{e}_{i_1}), \ldots, \beta(\underline{e}_{i_s})$ span $E^s = E''$ at the point x. But it is easily checked that the condition $\beta(\underline{e}_{i_1}), \ldots, \underline{e}_{i_s})$ span E^s is an open condition; hence there exists a (possibly) smaller neighbourhood $V_x \subset U_x$ in which this is the case. In V_x we may therefore define a map $\sigma_x : E'' \longrightarrow E$ by $\sigma_x\left(\beta\left(\underline{e}_{i_j}\right)\right) = \underline{e}_{i_j}$ for $j = 1, \ldots, s$. Since X is paracompact, there exists a partition of unity $\{\phi_x\}_{x \in X}$ associated to $\{V_x\}_{x \in X}$.

Now if we set $\sigma = \sum_{x \in X} \sigma_x \phi_x$, the sum makes sense by property iv) of partition of unity, and is continuous by property ii), and defines a splitting of $\beta : E \longrightarrow E''$ by property iii), since

$$\beta \cdot \sigma = \sum_x (\beta \cdot \sigma_x) \phi_x = \sum_x (\mathrm{Id}_{E''}|_{V_x}) \phi_x = \mathrm{Id}_{E''} \ .$$

ii) Since E is locally trivial we know that there is an open cover $\mathcal{U}$ of X such that for each $U \in \mathcal{U}$, $E|_U \approx E_U^n$ i.e. there exist n sections $f_1^U, \ldots, f_n^U \in \Gamma(U, E)$ which span E. Since X is compact, there exists a finite subcover $\mathcal{V}$ of $\mathcal{U}$, and a partition of unity $\{\rho_v\}_{v \in \mathcal{V}}$ supported in V. Then we can define a map

$$f : \oplus_{v \in \mathcal{V}} E^n \longrightarrow E$$

with f restricted to the "v-th" factor equal to $\left(\rho_V f_1^V, \ldots, \rho_V f_n^V\right)$. It is easily checked that f is surjective, and since $\#(\mathcal{V})$ is finite , we have constructed a surjective map $\beta : E^m \longrightarrow E$ for some m, which by i) must split, yielding the desired isomorphism $E^m \simeq E \oplus \ker(\beta)$. $\qquad\square$

2.4 K-theory of compact spaces

It follows from part ii) of theorem 2.3 that if X is a compact space, and E is
a vector bundle over X, then $\Gamma(X, E)$ is a finitely generated projective $\mathcal{C}(X)$-
module. In fact we have a theorem of Swan ([Sw], [Se3]):

Theorem . *If X is a compact space, then the global sections functor from vector
bundles over X to finitely generated projective $\mathcal{C}(X)$-modules is an equivalence
of categories.*

Let us briefly sketch the proof of this theorem.

It follows from theorem 2.3 that Γ is fully faithful; hence it suffices to show
that every projective $\mathcal{C}(X)$-module of finite rank is isomorphic to $\Gamma(X, E)$ for
some bundle E. If P is a finitely generated projective $\mathcal{C}(X)$-module, then P is
the image of a projector $P : \mathcal{C}(X)^n \longrightarrow \mathcal{C}(X)^n$ for some n. The projector induces
a map of bundles $\tilde{p} : E_X^n \longrightarrow E_X^n$ since Γ is fully faithful. It is easily checked
that $\mathrm{Im}(\tilde{p}) = \ker(1 - \tilde{p})$ is a subbundle $\tilde{P}$ of E^n, and that $\Gamma(X, \tilde{P}) = P$.

The construction of the bundle $\tilde{P}$ associated to a projective $\mathcal{C}(X)$-module can
be made more canonical. If X is a compact space, there is a $1 - 1$ correspondence
between the points of X and the maximal ideals of $\mathcal{C}(X)$

$$x \longmapsto m_x = \{f \in \mathcal{C}(X) \mid f(x) = 0\} \ .$$

Then if P is a projective $\mathcal{C}(X)$-module, the associated vector bundle, as a set, is

$$\tilde{P} = \bigcup_{x \in X} P \otimes_{\mathcal{C}(X)} (\mathcal{C}(X)/m_x \simeq \mathbb{C}) \ .$$

Exercise . Construct the topology on $\tilde{P}$.

An important consequence of the Serre-Swan theorem is that $K^0(X) =
K_0\big(\mathcal{C}(X)\big)$.

For more details of the Serre-Swan theorem, see [Ka1], [Sw], [Se1]. This theorem is the starting point for the work of Brown, Douglas and Filmore, and
others, on K-theory and operator algebras. See [B-D-F], [Dou].

Let us now consider the basic properties of $K^0(X) = K^0_{\mathrm{top}}(X)$ for X a
compact topological space. See [At], chapters 1 and 2 for details.

1. Functoriality. If $f : X \longrightarrow Y$ is continuous map between compact spaces,
then $f^* : \mathrm{Vect}_Y \longrightarrow \mathrm{Vect}_X$ is an exact functor (equivalent to the map $\mathcal{P}_{\mathcal{C}(Y)} \longrightarrow
\mathcal{P}_{\mathcal{C}(X)}$ given by $P \longrightarrow \mathcal{C}(X) \otimes_{\mathcal{C}(Y)} P$, and if $g : Y \longrightarrow Z$ is another such
map, there is an isomorphism of functors $f^* \cdot g^* = (g \cdot f)^*$. Hence we have a
contravariant functor from spaces to abelian groups, $K^0_{\mathrm{top}} : X \longmapsto K^0_{\mathrm{top}}(X) =
K_0(\mathrm{Vect}_X)$.

2. Products. Tensor product of vector bundles makes $K^0_{\mathrm{top}}(X)$ a ring: $[E] \cdot [F] =
[E \otimes F]$, (this corresponds to tensor product of $\mathcal{C}(X)$-modules). The product
structure is compatible with pull-backs.

3. Homotopy. If f, g are homotopic maps from X to Y, then $f^* = g^* : K_0(Y) \longrightarrow K_0(X)$.

4. Representability. We saw in (2.3) that if X is compact and E is a vector bundle (of finite rank r) over X, then E is the quotient of a trivial bundle $\mathbb{C}^n \times X$. Hence for each $x \in X$ the fibre E_x of E over x is a quotient of $\mathbb{C}^n$, thus defines a point in the Grassmannian $G(r, n)$ of r-dimensional quotients of $\mathbb{C}^n$. One can show easily that this point varies continuously with $x \in X$. That is, there is a continuous map

$$\phi_E : X \longrightarrow G(r, n)$$

such that $E \simeq \phi_E^*(U_{r,n})$, where $U_{r,n}$ is the universal rank r quotient of $\mathbb{C}^n$ over $U_{r,n}$. Then ([At] Theorem 1.4.15) there is a bijection between the set of isomorphism classes of rank r bundles over X and $\varinjlim_n [X, G(r, n)]$. Here $[,]$ denotes homotopy classes of maps, and $G(r, n) \longrightarrow G(r, n + 1)$ is induced by the projection $\mathbb{C}^{n+1} \simeq \mathbb{C}^n \times \mathbb{C} \longrightarrow \mathbb{C}^n$.

Now the direct limit of Grassmannians $\varinjlim_n G(r, n)$ is a classifying space $BU(n)$ for the unitary group $U(n)$. Taking the limit as $n \to \infty$, we get BU, the classifying space of the infinite unitary group, and an isomorphism

$$K^0_{\text{top}}(X) \cong [X, \mathbb{Z} \times BU]$$

(the projection $[X, \mathbb{Z} \times BU] \longrightarrow [X, Z] = \mathbb{Z}^X$ is the rank homomorphism).

5. Half-Exactness. If $(X, Y), Y \subset X$ is a compact pair, we define $K^0_{\text{top}}(X, Y) = \tilde{K}^0_{\text{top}}(X/Y)$ where $\tilde{K}^0_{\text{top}}(T)$ for T a space with basepoint $t \in T$ is the kernel of $K_0(T) \longrightarrow K_0(t) = \mathbb{Z}$. If $j : X \longrightarrow X/Y$ is the projection and $i : Y \longrightarrow X$ the inclusion, there is an exact sequence

$$K^0_{\text{top}}(X, Y) \xrightarrow{j^*} K^0_{\text{top}}(X) \xrightarrow{i^*} K^0_{\text{top}}(Y) \ .$$

2.5. Higher K-Theory

The basic properties 1–5 listed in 2.4 are properties of a generalized cohomology theory ([Dy], [Ad]) i.e. they are all the properties of singular cohomology except that the dimension axiom is missing. In particular we can define groups $K^{-n}_{\text{top}}(X)$ as follows. Let $X^+ = X \cup \{\phi\}$ be X with a disjoint basepoint; then $S^n X^+ = S^n \times (X^+)/(* \times X^+ \cup S^n \times \{\phi\})$ is the n-fold reduced suspension of X^+. We define

$$K^{-n}_{\text{top}}(X) = \tilde{K}^0_{\text{top}}\left(S^n(X^+)\right) \ .$$

Here $\tilde{K}^0_{\text{top}}(X) = K^0_{\text{top}}(X, *)$ if X is a compact space with basepoint $*$. Using the half-exactness of K_0, one can show that, for all pairs (Y, X), there is a long exact sequence:

$$\ldots \longrightarrow K^{-n}(Y) \xrightarrow{\delta} \tilde{K}^{-n+1}(X/Y) \xrightarrow{j^*} \tilde{K}^{-n+1}(X) \xrightarrow{i^*} \tilde{K}^{-n+1}(Y) \longrightarrow \ldots \ .$$

Using the definition of K^{-n}, together with the representability of K^0, one can see that for $n \geq 1$:

$$K^{-n}(X) \simeq [X, \Omega^n BU]$$
$$\simeq \pi_n(BU^X) = \pi_n(B(U^X))$$

where we give BU^X the compact open topology and U^X is the group of continuous maps from X to U, again with the compact open topology.

An important feature of topological K-theory is *Bott periodicity*, i.e. that there is a natural isomorphism $K^0(X) \longrightarrow K^{-2}(X)$. Since K-theory is representable, periodicity may be expressed as a homotopy equivalence

$$\Omega U = \mathbb{Z} \times BU$$

(Recall that $\Omega BU \simeq U$). Also taking X to be a point, we get

$$K^{-n}(pt) = \pi_n(\mathbb{Z} \times BU) = \begin{cases} \mathbb{Z} & n \text{ even} \\ 0 & n \text{ odd} \end{cases} .$$

Bott periodicity is one of the fundamental properties of topological K-theory, and we shall later be interested in the question of how close algebraic K-theory comes to being periodic.

2.6. The Atiyah-Hirzebruch Spectral Sequence

One consequence of the Bott periodicity theorem and the resulting computation of $K_*(pt)$ is that we can relate K-theory and singular cohomology with integer coefficients. There is a spectral sequence constructed by Atiyah and Hirzebruch:

$$E_2^{p,q} = H^p(X, K^q(pt)) = \begin{cases} H^p(X, \mathbb{Z}) & q \text{ even}, \ \ q \leq 0 \\ 0 & \text{otherwise} \end{cases} \Rightarrow K^{p+q}(X) .$$

We shall see later that, after tensoring with $\mathbb{Q}$, this spectral sequence degenerates at E_2, giving an isomorphism

$$K^n(X) \otimes_{\mathbb{Z}} \mathbb{Q} \simeq \bigoplus_{\substack{p+2r=n \\ r \leq 0}} H^p(X, \mathbb{Q}) .$$

2.7. Chern Classes

A key set of tools in studying K-theory, both topological and algebraic, is the theory of characteristic classes. These are natural transformations from K-theory to other (usually more familiar) cohomology theories. We shall be most interested in Chern classes for complex vector bundles.

If $\pi : E \longrightarrow X$ is a complex vector bundle over a paracompact topological space, there are Chern Classes $c_i(E) \in H^{2i}(X, \mathbb{Z})$ which are characterized by:

0) $c_0(E) = 1 \in H^0(X, \mathbb{Z})$
1) $c_n(E \oplus F) = \sum_{i+j=n} c_i(E) \cup c_j(F)$ for all $n \geq 0$

2) If $f : Y \longrightarrow X$ is a continuous map, then $c_n(f^*E) = f^*c_n(E)$, for all $n \geq 0$.

3) Let $P^n(\mathbb{C})$ be the space of rank 1 quotients of $\mathbb{C}^{n+1}$, we shall write ξ_n for the tautological line bundle on $P^n(\mathbb{C})$, which sits in an exact sequence:

$$0 \longrightarrow H \longrightarrow E^{n+1} \longrightarrow \xi_n \longrightarrow 0 \ .$$

As is well known $H^2\big(\mathbb{P}^n(\mathbb{C}), \mathbb{Z}\big)$ is infinite cyclic with generator h_n, Poincaré dual to the hyperplane $\{z_n = 0\}$, identified with $\mathbb{P}^{n-1}(\mathbb{C})$ with the natural orientation. Then

$$c_i(\xi_n) = \begin{cases} h_n & \text{if } i = 1 \\ 0 & \text{if } i > 0 \end{cases} \ .$$

Axiom 1), the Whitney sum formula, implies that $c(E) = \sum_{i>0} c_i(E) \in H^*(X, \mathbb{Z})$ depends only on the class of the vector bundle E in $K_0(X)$, and that the Chern classes extend to functions $c_i : K_0(X) \longrightarrow H^{2i}(X, \mathbb{Z})$. We may define a new characteristic class, the *Chern character* ch $: K_0(X) \longrightarrow H^*(X, \mathbb{Q})$, satisfying the axioms:

0) $\text{ch}_0(E) = \text{rk}(E) \in H^0(X, \mathbb{Z}) \simeq \mathbb{Z}$

1) $\text{ch}(E \oplus F) = \text{ch}(E) + \text{ch}(F)$

2) $\text{ch}(f^*E) = f^*\text{ch}(E)$ if $f : Y \longrightarrow X$ is a continuous map.

3) $\text{ch}(\xi_n) = \exp(h_n) \in H^*(P^n(\mathbb{C}), \mathbb{Q})$.

Then one may check that ch extends to a ring homomorphism

$$\text{ch} : K^0_{\text{top}}(X) \longrightarrow H^*(X, \mathbb{Q})$$

and that $\text{ch}_1(E) = c_1(E)$, and $\text{ch}_i(E)$ is a polynomial in $\text{rk}(E), c_1(E), \ldots, c_i(E)$ with rational coefficients for all i.

Of course we have not checked the existence or uniqueness of Chern classes, or the Chern character; a good reference for this is [Hi] or [Mi-S].

3. K_0 of Schemes

3.1. Affine Schemes

We saw in (2.4) that the category of complex vector bundles over a compact topological space X is equivalent to the category of projective $\mathcal{C}(X)$-modules. Furthermore the space X itself can be recovered as the set of maximal ideals in $\mathcal{C}(X)$. We shall sketch the analogous theory for algebraic varieties.

If R is a ring (commutative, with unit), we can form a topological space $\text{Spec}(R)$ with points the set of prime ideals on R and the topology given by saying that a set $Y \subset \text{Spec}(R)$ is closed if and only if there is an ideal $I \triangleleft R$ for which Y is the set of prime ideals containing I. We write $Y = V(I)$, and viewing R as the ring of functions on $\text{Spec}(R)$ we may view $V(I)$ as the set of points $y \in \text{Spec}(R)$ at which all the functions in I vanish. When we view $f \in \text{Spec}(R)$ as a function on $\text{Spec}(R)$, its "value" at $y \in \text{Spec}(R)$ is its image in the residue field $k(y)$ (i.e. the field of fractions of the integral domain R/y).

If $X = \text{Spec}(R)$, then there is a sheaf of rings, $\mathcal{O}_X$ on X, analogous to the sheaf of continuous functions on a topological space, with $\mathcal{O}_X(U) = S^{-1}R$, where $S = \{f \in R \mid V, y \in U, f \notin y\}$. (Note $f \notin y$ can be interpreted as "f does not vanish at y".) In particular $\mathcal{O}_X(X) = R$. The ringed space $X = (X, \mathcal{O}_X)$ is called an *affine scheme*.

3.2. Schemes

A *scheme* is a topological space X together with sheaf of rings $\mathcal{O}_X$ on X such that X has a covering by open subsets $U \subset X$, such that each $(U, \mathcal{O}_X |_U)$ is isomorphic to an affine scheme.

A morphism $f : X \longrightarrow Y$ of schemes is a continuous map of the underlying topological spaces, together with for each open set $U \subset Y$ a ring homomorphism

$$f_U^* : \mathcal{O}_Y(U) \longrightarrow \mathcal{O}_X\left(f^{-1}(U)\right)$$

which is compatible with the obvious restriction maps for each inclusion $V \subset U$. In addition we require that if $x \in f^{-1}(U)$ and $g \in \mathcal{O}_Y(U)$, then if g vanishes at $f(x)$, $f^*(g) \in \mathcal{O}_X\left(f^{-1}(U)\right)$ vanishes at x. Here we say that $f \in \mathcal{O}_Y(U)$ vanishes at a point $y \in U$ if given any affine neighbourhood W of y, the image of f in $\mathcal{O}_W(U \cap W)$ lies in the prime ideal corresponding to y.

Here we are following the more recent convention of, for example [Ha]. Earlier texts called the object defined above a pre-scheme and schemes were required to be separated.

An important fact is that affine schemes form a full subcategory of the category of schemes, which is isomorphic to the category of commutative rings, except that the arrows are reversed; i.e. $j : R \longrightarrow S$ corresponds to $\operatorname{Spec}(f) : \operatorname{Spec}(S) \longrightarrow \operatorname{Spec}(R)$.

Example . Let R be a ring. Then we define a scheme $\mathbb{P}_R^n$, projective n-space over R, as follows. $\mathbb{P}_R^n$ is the union of $n + 1$ affine open sets $U_0, \ldots, U_n$ each isomorphic to the affine scheme $\operatorname{Spec}(R[T_1, \ldots, T_n])$. Let $X_0, \ldots, X_n$ be a set of $n + 1$ variables, and for each $i = 1, \ldots, n$ let

$$U_i = \operatorname{Spec} R[X_0/X_i, \ldots, X_{i-1}/X_i, X_{i+1}/X_i, \ldots, X_n/X_i]$$

(we may view the coordinate ring of U_i as a subring of the localization of $R[X_0, \ldots, X_n]$ with respect to the multiplicative system $\{X_i^k\}_{k \in \mathbf{Z}^+}$). If $i \neq j$, then $U_i \cap U_j = \operatorname{Spec}\left(R\left[X_0/X_i, \ldots, X_n/X_i, (X_j/X_i)^{-1}\right]\right)$ which is an affine subset of both U_i and U_j.

A feature of the category of schemes is the existence of fibre products. If $f : X \longrightarrow Z$ and $g : Y \longrightarrow Z$ are morphisms there is a scheme $X \times_Z Y$ and morphisms $p : X \times_Z Y \longrightarrow X$ and $q : X \times_Z Y \longrightarrow Y$ for which $f \cdot p = g \cdot q$, such that if $p' : U \longrightarrow X, q' : U \longrightarrow Y$ is any other pair of morphisms with the same property, there is a unique map $h : U \longrightarrow X \times_Z Y$ satisfying $p' = p \cdot h$ and $q' = q \cdot h$:

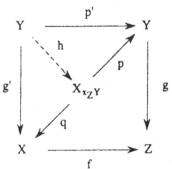

If $X = \mathrm{Spec}(A), Y = \mathrm{Spec}(B)$ and $Z = \mathrm{Spec}(C)$ are all affine, then $X \times_Z Y = \mathrm{Spec}\,(A \otimes_C B)$. From this example one may see that $X \times_Z Y$ will not be the fibre product of X and Y as sets or topological spaces.

If X is a scheme and $S = \mathrm{Spec}(R)$ is affine, we write $X(R)$ for the set of morphism of schemes $S \longrightarrow X$; this is called the set of R valued point of X. For example, for affine n-space over $\mathbb{Z}, \mathbf{A}_{\mathbb{Z}}^n = \mathrm{Spec}\,(\mathbb{Z}[t_1, \cdots t_n])$, an R-valued point of $\mathbf{A}_{\mathbb{Z}}^n$ is given by a ring homomorphism $\mathbb{Z}[t_1, \cdots t_n] \longrightarrow R$, i.e. by an n-tuple $r_1, \cdots, r_n$ of elements of R. For a non affine example, one may easily check that $\mathbb{P}_{\mathbb{Z}}^n(\mathbb{C}) = \mathbb{P}(\mathbb{C})$. Here one sees a special case of a general phenomenon. Let X be a scheme over a subring $R \subset \mathbb{C}$ (i.e. there is a morphism $\pi : X \longrightarrow \mathrm{Spec}(R)$). Then the set of $\mathbb{C}$-valued points $p : \mathrm{Spec}\,\mathbb{C} \longrightarrow X$ such that $p^* \cdot \pi^* : R \longrightarrow \mathbb{C}$ is the inclusion will, under mild hypotheses on the map π naturally be the set of points of a complex analytic space, or even a complex manifold. Finally, note that every scheme X admits a unique map $\pi : X \longrightarrow \mathrm{Spec}\,\mathbb{Z}$; if X is a scheme we shall usually write $\mathbf{A}_X^n$ for $\mathbf{A}_{\mathbb{Z}}^n \times_{J\,\mathrm{Spec}\,\mathbb{Z}} X$.

3.3. Algebraic bundles

If X is a scheme we can define a "geometric" vector bundle in a fashion analogous to the definition of §2 for the topological case. That is, a morphism of schemes $\pi : E \longrightarrow X$ together with maps

$$S : E \times_X E \longrightarrow E$$

$$\mu : \mathbf{A}_{\mathbb{Z}}^1 \times_{\mathrm{Spec}(\mathbb{Z})} \mathbf{E} \longrightarrow \mathbf{E}$$

satisfying rules analogous to i), ii), iii) ibid, together with the local triviality condition that there is an open cover $\{U_\alpha\}$ of X together with isomorphisms $E \mid_{U_\alpha} \approx A_{U_\alpha}^n$ – for each α.

However, the most common way of viewing a vector bundle $\pi : E \longrightarrow X$ over X is via its sheaf of sections $\mathcal{E}$:

$$\mathcal{E}(U) = \text{maps } s : U \longrightarrow E \quad \text{such that} \quad \pi \circ s = \mathrm{Id}_U .$$

$\mathcal{E}$ is then required to be a *locally free* sheaf of $\mathcal{O}_X$ modules i.e. there is an open cover $\{U_\alpha\}$ of X such that $\mathcal{E}\mid_{U_\alpha} \approx \mathbf{A}_{U_\alpha}^{n_\alpha}$ for each α, with $n_\alpha \in \mathbf{N}$.

Remark. In general there are two ways of associating a vector bundle E to a locally free sheaf $\mathcal{E}$

i) so that $\Gamma(E) \simeq \mathcal{E}$

ii) so that $\Gamma(E^*) \simeq \mathcal{E}$, where E^* is the *dual* of E.

We are using i); however in algebraic geometry, following Grothendieck, one usually uses approach ii). Vector bundles over X form a category, in which a morphism $f : E \longrightarrow F$ may be described either in terms of geometric bundles, analogously to §2, or in terms of the sheaves of sections $\mathcal{E}$, $\mathcal{F}$, of respectively E and F. A homomorphism of bundles is just an $\mathcal{O}_X$ linear map $f : \mathcal{E} \longrightarrow \mathcal{F}$; i.e. for each open subset $U \subset X$, an $\mathcal{O}_X(U)$ linear map of modules, $f_U : \mathcal{E}(U) \longrightarrow \mathcal{F}(U)$, such that, if $C \subset U$, then $\rho_{VU}\, f_U = f_V\, \rho_{VU}$. We shall say that a sequence

$$0 \longrightarrow E \longrightarrow F \longrightarrow G \longrightarrow 0$$

is *exact* if the corresponding sequence

$$0 \longrightarrow \mathcal{E}(U) \longrightarrow \mathcal{F}(U) \longrightarrow \mathcal{G}(U) \longrightarrow 0$$

is exact for every *affine* open set $U \subset X$.

If X is affine, i.e. $X = \mathrm{Spec}(R)$ for R a ring (when X is a variety over a field k, this means $X \subset \mathbf{A}_k^n$), then every exact sequence of vector bundles on X splits. In fact the functor $E \longrightarrow \Gamma(X, \mathcal{E})$ defines an equivalence of categories between the category of vector bundles on X and the category of projective R-modules. However if X is not affine, then in general, an exact sequence of bundles over X will *not* split.

Example . In general, if $\mathcal{F}$ is a sheaf of $\mathcal{O}_X$-modules, then the first sheaf cohomology group ([Ha], [Go]) $H^1(X, \mathcal{F})$ is isomorphic to the Yoneda Ext group $\mathrm{Ext}_{\mathcal{O}_X}^1(\mathcal{O}_X, \mathcal{F})$. So if $H^1(X, \mathcal{F}) \neq 0$, there is an exact sequence of locally free sheaves $0 \longrightarrow \mathcal{F} \longrightarrow \mathcal{G} \longrightarrow \mathcal{O}_X \longrightarrow 0$, or equivalently a sequence $0 \longrightarrow F \longrightarrow G \longrightarrow E^1 \longrightarrow 0$ of bundles, which does *not* split.

In particular Serre computed in [Se1] that $H^1(\mathbf{P}^n, \mathcal{O}(n)) \neq 0$ where $\mathcal{O}(n) = \mathcal{O}(1)^{\otimes n}$ and $\mathcal{O}(1)$ is the sheaf of (algebraic) sections of the tautological line bundle ξ on $\mathbf{P}^n$ introduced in §2.7. (ξ is easily seen to be an algebraic bundle).

The difference between the algebraic and topological cases is that algebraic partitions of unity do not exist. If an algebraic function vanishes on an open set of an irreducible scheme (i.e. a scheme which is not the union of two proper closed subsets), then it vanishes everywhere. Hence in the algebraic case we cannot pass from the existence of local splittings of an exact sequence to the existence of a global splitting.

As well as algebraic bundles over schemes we can consider holomorphic bundles over complex manifolds. If a compact manifold M is in fact algebraic (i.e.

it is isomorphic to the set of $\mathbb{C}$-valued points of some scheme over $\mathbb{C}$ – which is always the case if M is a complex submanifold of $P^n_{\mathbb{C}}$), then every holomorphic bundle on M is algebraic (Serre, [Se2]).

3.4. K_0 of Schemes

If X is a scheme, we now define $K_0(X)$ to be the Grothendieck group $K_0(\mathcal{P}(X))$ of the exact category $\mathcal{P}(X)$ of vector bundles, or equivalently, locally free sheaves on X. Given the remarks above, we must use exact sequences to give the relations. If E is a vector bundle, and $\mathcal{E}$ the corresponding locally free sheaf of sections of E, we write $[E] = [\mathcal{E}]$ for the corresponding element of $K_0(X)$.

Just as for rings, we have a product structure on $K_0(X)$ given by

$$[\mathcal{E}] \cdot [\mathcal{F}] = [\mathcal{E} \otimes_{\mathcal{O}_X} \mathcal{F}]$$

for vector bundles over X; here if $U \subset X$, is an affine open set

$$\mathcal{E} \otimes \mathcal{F}(U) = \mathcal{E}(U) \otimes_{\mathcal{O}_X} \mathcal{F}(U) .$$

This product makes $K_0(X)$ a commutative ring.

If $f : X \longrightarrow Y$ is a morphism of schemes, there is an exact functor

$$f^* : \mathcal{P}(Y) \longrightarrow \mathcal{P}(X)$$
$$\mathcal{E} \longrightarrow f^*\mathcal{E} .$$

Here if $U \subset X$, $V \subset Y$ are affine open sets with $f(U) \subset V$,

$$f^*\mathcal{E}(U) = \mathcal{O}_X(U) \otimes_{\mathcal{O}_Y(V)} \mathcal{E}(V) .$$

This functor is easily seen to be exact, and to induce a ring homomorphism

$$K_0(f) = f^* : K_0(Y) \longrightarrow K_0(X) .$$

It is straight forward to verify that K_0 is now a *contravariant* functor from the category of schemes to the category of commutative rings. Composing this K_0 with the contravariant fully faithful functor "Spec" from rings to schemes gives us back the *covariant* functor K_0 on rings.

Examples

(i) X is a smooth projective curve over a field k; (see [Ha] for what this means) then

$$K_0(X) \simeq \mathbb{Z} \oplus \mathrm{Pic}(X)$$
$$[E] \longrightarrow \mathrm{rk}(E) \oplus \left[\Lambda^{\mathrm{rk}(E)} E \right]$$

where $\mathrm{Pic}(X)$ is the group of isomorphism classes of line bundles (i.e. vector bundles of rank 1) over X. If $k = \mathbb{C}$, then $\mathrm{Pic}(X) \simeq \mathbb{Z} \oplus \mathbb{C}^g/\Lambda$ with $\mathbb{C}^g/\Lambda$ a compact complex torus equal to the quotient of $\mathbb{C}^g$ by a lattice Λ. See [G-H] Chapter 2.6 for details.

(ii) Let X be a noetherian separated schema, $\pi : \mathbb{P}^n_X = \mathbb{P}^n_{\mathbf{Z}} \times_{\mathrm{Spec}_{\mathbf{Z}}} X \longrightarrow X$ the natural map.

Let ξ_n be the tautological line bundle on $\mathbb{P}^n_X$, i.e. the pull back of the tautological line bundle on $\mathbb{P}^n_{\mathbf{Z}}$. Write $x = [\xi_n] - [\mathcal{O}_X]$. Then the map

$$K_0(X)^{\oplus(n+1)} \longrightarrow K_0(\mathbb{P}^n_X)$$

$$\alpha \longmapsto \sum_{i=0}^{n} \pi^*(\alpha) \cdot x^i$$

is an isomorphism. In particular, if $X = \mathrm{Spec}(k)$, k a field,

$$K_0(\mathbb{P}^n_k) \simeq \mathbf{Z}^{n+1} \ .$$

For a proof see [SGA6] VI 1.1 and [Q2] §8. Notice that examples (i) and (ii) intersect in the case $X = \mathbb{P}^1_k$, in which case

$$K_0(\mathbb{P}^1_k) \simeq \mathbf{Z} \oplus \mathbf{Z}$$

$$[E] \longmapsto \mathrm{rk}(E) \oplus \deg\left(\Lambda^{\mathrm{rk}(E)} E\right) \ .$$

3.5. K_0 and Algebraic Cycles

In the previous section we saw that if X is a smooth projective curve over a field, then $K_0(X) = \mathbf{Z} \oplus \mathrm{Pic}(X)$. As is well known (see for example [Ha] Ch II section 6), $\mathrm{Pic}(X)$ is isomorphic to the group of divisors on X modulo the subgroup of principal divisors. Here a divisor is an element $\sum_{P \in X} n_P[P]$ of the free abelian group on the closed points of X, and a divisor D is *principal* if there is a non-zero rational function f on X such that $D = \sum_{P \in X} v_P(f)[P]$ where $v_P(f)$ is the order of vanishing of f at the point P. We write $D = (f)$ or $D = \mathrm{div}(f)$. The isomorphism is given by

$$\mathcal{L} \in \mathrm{Pic}(X) \longrightarrow \mathrm{div}(s) = \Sigma v_P(s)[P]$$

where s is any meromorphic section of $\mathcal{L}$, i.e. element of the stalk of $\mathcal{L}$ at the generic point of X, and $v_P(s)$ is the order of vanishing of s at the closed point in X. (Note that in both cases, $v_P(s)$ or $v_P(f)$ can take negative values if there is a pole at P.)

This result may be generalized for general schemes as follows. If X is a scheme, $Z^p(X)$ is the free abelian group on reduced irreducible subschemes $Z \subset X$ of codimension p (i.e. the Krull dimension of the local ring $\mathcal{O}_{X,z}$ at the generic point z of Z is p.) We say $\Sigma n_i[Z_i] \in Z^p(X)$ is *rationally equivalent to zero* if there is a finite collection of reduced irreducible subschemes W_j of codimension $p - 1$, and rational functions $f_j \in k(W_j)$ such that

$$\Sigma \mathrm{div}(f_j) = \Sigma n_i[Z_i] \ .$$

Here $\mathrm{div}(f)$, for f a rational function on a codimension $p-1$ reduced irreducible subscheme, is $\Sigma\, v_Z(f)[Z]$ where $v_Z(f)$ is the order of vanishing of f at the generic point of Z, and the sum is taken over all codimension p reduced irreducible subvarieties. We write $CH^p(X)$ for the resulting quotient group.

For X a regular scheme, and especially for X a smooth variety over a field, one may view X as being somewhat analogous to the singular cohomology of X. In particular the graded group $\oplus_{p \geq 0} CH^p(X)$ has a natural product structure.

Extending the result for curves, and by analogy with the isomorphism in the topological category $K^0_{\mathrm{top}}(X)_{\mathbb{Q}} \simeq H^{2*}(X, \mathbb{Q})$, there is an isomorphism of rings, for X a smooth variety over a field:

$$K_0(X)_{\mathbb{Q}} = \bigoplus_{p \geq 0} CH^p(X)_{\mathbb{Q}} \ .$$

See [SGA6], exp XIV and [So].

4. Higher K-Theory of Varieties and Schemes

4.1

A natural question, on seeing the close analogy between algebraic and topological K_0, in particular the Serre-Swan theorem, is whether there are 'higher' *algebraic* K-theory groups which are analogous to the higher topological K-theory groups. A number of constructions of such groups have been given, with the two most important being the "Q" and "$+$" constructions of Quillen. (See [An] for a general survey of constructions of higher K-theory.)

We need to start by recalling some basic properties of simplicial sets and nerves of categories.

Let us write Δ for the category of finite non-empty totally ordered sets. If $\mathcal{C}$ is a category, a *simplicial object* $X.$ in $\mathcal{C}$ is a *contravariant* functor $X. : \Delta \longrightarrow \mathcal{C}$. Such a functor is totally determined by the sequence of objects $X_n = X(\{0 < \ldots < n\})$ in $\mathcal{C}$, together with the 'face' and 'degeneracy' maps $d_i : X_n \longrightarrow X_{n-1}$, $s_i : X_{n-1} \longrightarrow X_n$ induced by the maps $\delta_i : [n-1] = \{0 < \ldots < n-1\} \longrightarrow [n] = \{0, \ldots, n\}$ and $\sigma_i : [n] = \{0 < \ldots < n\} \longrightarrow [n-1]\{0 < \ldots < n-1\}$, where δ_i, for $0 \leq i \leq n$, is the unique injective monotone map such that i is *not* in the image of δ_i, while σ_i is the unique surjective monotone map with $\sigma_i(i) = \sigma_i(i+1)$, for $0 \leq i \leq n-1$.

If $X.$ is a simplicial set, i.e. a simplicial object in the category of sets, it has a *geometric realization* $|X.|$, which is a CW complex, the p-cells of which are in $1-1$ correspondence with the non-degenerate elements of X_p, i.e. with $X_p - \bigcup_{i=0}^{p-1} s_i(X_{p-1})$. The attaching maps for the CW structure are determined by the face maps $d_i : X_p \longrightarrow X_{p-1}$. The geometric realization is natural, so that a morphism $f : X. \longrightarrow Y.$ (i.e. a natural transformation of functors) induces a continuous map $|f| : |X.| \longrightarrow |Y.|$.

Examples

1. Nerves of Categories

We may regard the elements of any finite totally ordered set A as the objects of a category in which there is one morphism $a \longrightarrow b$ if $a \leq b$ and none otherwise. Any monotonic map $f : A \longrightarrow B$ may then be viewed as a functor.

Now suppose that $\mathcal{C}$ is a small category, i.e. a category in which the objects form a set. We define a simplicial set $N.\mathcal{C}$, the *nerve* of $\mathcal{C}$, as follows. For each finite totally ordered set, the value of $N.\mathcal{C}$ on A, $N_A\mathcal{C}$, is the set of functors $A \longrightarrow \mathcal{C}$. Hence if $A = \{0, \ldots, k\}$, $N_k\mathcal{C}$ is the set of k-tuples of composable arrows, $C_0 \longrightarrow C_1 \longrightarrow \ldots \longrightarrow C_k$ in $\mathcal{C}$. The face and degeneracy maps are then determined by the *functors* δ_i and σ_i. Finally, we write $B\mathcal{C}$ for the geometric realization of $N.\mathcal{C}$, and call it the *classifying space* of $\mathcal{C}$.

2. Classifying Spaces of Groups

If G is a (discrete) group, we may view the elements of G as the morphisms in a category with one object. Then we set BG equal to the classifying space of this category. One can show that $\pi_1 BG = G$, and $\pi_i BG = 0$ for $i > 1$.

4.2. The Q-construction

Returning to the construction of higher K-theory, we start by describing Quillen's Q-construction ([Q2] §2).

Definition . Let $\mathcal{M}$ be an exact category. We define a new category $Q\mathcal{M}$, with the same objects as $\mathcal{M}$, as follows. If M and M' are objects in $\mathcal{M}$, a morphism from M to M' in $Q\mathcal{M}$ consists of an isomorphism class of diagrams in $\mathcal{M}$ of the form:

$$M \xleftarrow{j} N \xrightarrowtail{i} M'$$

with i an admissible monomorphism, and j an admissible epimorphism. Composition is defined as follows; if $M \xleftarrow{j} N \xrightarrowtail{i} M'$ and $M' \xleftarrow{j'} N' \xrightarrowtail{i'} M''$ are morphisms, let $P = N \times_{M'} N'$, so that we have a diagram:

$$
\begin{array}{ccccc}
P & \rightarrowtail & N' & \rightarrowtail & M'' \\
\downarrow & & \downarrow & & \\
N & \rightarrowtail & M' & & \\
\downarrow & & & & \\
M & & & &
\end{array}
$$

Then define the composition to be the isomorphism class of $M \twoheadleftarrow P \longrightarrow M''$.

It is easy to check that if $F : \mathcal{M}_1 \longrightarrow \mathcal{M}_2$ is an exact functor (i.e. an additive functor preserving exact sequences) then there is an induced functor $Q(F) : Q\mathcal{M}_1 \longrightarrow Q\mathcal{M}_2$.

Definition . If $\mathcal{M}$ is a small exact category, we set $K_i(\mathcal{M}) = \pi_{i+1} BQ\mathcal{M}$ for $i \geq 0$. One can show that for $i = 0$ this definition agrees with the previous definition in §1.

Properties of K_*M. We now list some of the properties of this definition of K-theory. For details and more properties see [Q2].

1. Functoriality. Since an exact functor $F : \mathcal{M}_1 \longrightarrow \mathcal{M}_2$ induces a functor $QF : Q\mathcal{M}_1 \longrightarrow Q\mathcal{M}_2$, and hence a continuous map $BQ\mathcal{M}_1 \longrightarrow BQ\mathcal{M}_2$, it induces homomorphisms $K_*(F) : K_*(\mathcal{M}_1) \longrightarrow K_*(\mathcal{M}_2)$.

2. Products. If $\mathcal{M}_1, \mathcal{M}_2$ and $\mathcal{N}$ are exact categories, and $F : \mathcal{M}_1 \times \mathcal{M}_2 \longrightarrow \mathcal{N}$ is a biexact functor, i.e. $F(M_1, M_2)$ is exact in each variable separately, there is an induced bilinear product for $p, q \geq 0$:

$$K_p(\mathcal{M}_1) \times K_q(\mathcal{M}_2) \longrightarrow K_{p+q}(\mathcal{N}) \ .$$

3. Reduction by Resolution. If $\mathcal{M} \subset \mathcal{N}$ is a subcategory of an exact category such that:

1) $\mathcal{M}$ is closed under extensions in $\mathcal{N}$.
2) If $P \twoheadrightarrow M$ with P in $\mathcal{M}$ then M is in $\mathcal{M}$.
3) Every object N in $\mathcal{N}$ has a finite resolution $0 \longrightarrow P_n \longrightarrow P_{n-1} \longrightarrow \ldots \longrightarrow P_0 \longrightarrow N$ by objects in $\mathcal{M}$.

Then the inclusion $\mathcal{M} \subset \mathcal{N}$ induces an isomorphism $K_*(\mathcal{M}) \longrightarrow K_*(\mathcal{N})$.

Remark. One can show that $BQ\mathcal{M}$, for $\mathcal{M}$ an exact category, is homotopy equivalent to the geometric realization of the simplicial set $S.\mathcal{M}$ defined as follows. Given $n \geq 0$, $S_n\mathcal{M}$ is the set of diagrams

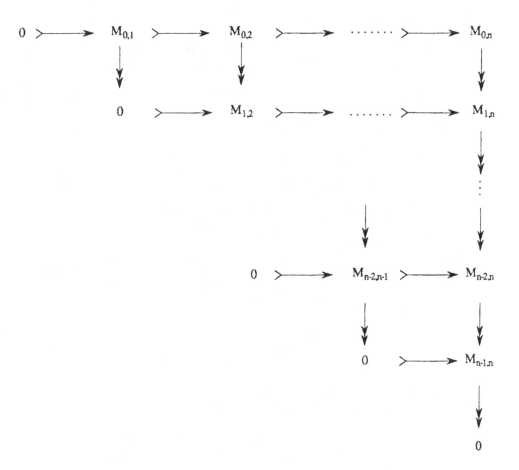

in which for $i \leq j \leq k$, $M_{ij} \rightarrowtail M_{ik} \longrightarrow M_{jk}$ is an admissible exact sequence. The face and degeneracy maps are defined in the obvious fashion, so that, for example, if $n = 2$,

$$d_0 (M_{01} \rightarrowtail M_{02} \twoheadrightarrow M_{12}) = M_{12}$$
$$d_2 (M_{01} \rightarrowtail M_{02} \twoheadrightarrow M_{12}) = M_{02}$$
$$d_2 (M_{01} \rightarrowtail M_{02} \twoheadrightarrow M_{12}) = M_{01}$$

Then $|S.\mathcal{M}|$ is essentially the classifying space for the addition law on objects in $\mathcal{M}$, given by $[M'] + [M''] = [M]$ if there is an admissible exact sequence $0 \longrightarrow M' \rightarrowtail M \twoheadrightarrow M'' \longrightarrow 0$. See [W], for more details. See also [G-G] for a construction of the loop space of the Q-construction.

4.3. K-theory of Rings and Schemes

Let X be a variety over a field, or more generally a (noetherian, separated) scheme. There are two categories of sheaves of $\mathcal{O}_X$-modules of interest:

$\mathcal{P}_X$ = The exact category of sheaves of $\mathcal{O}_X$-modules which are locally free of finite rank. The admissible exact sequences are the sequences which are exact in the usual sense. As in (2.4), $\mathcal{P}_X$ is equivalent to the category of vector bundles over X.
$\mathcal{M}_X$ = The category of *all* coherent $\mathcal{O}_X$-modules.

If $X = \operatorname{Spec}(R)$, R a (noetherian) ring, these categories are equivalent to:

$\mathcal{P}_X$ = the category of finitely generated ($f.g.$)projective R-modules.
$\mathcal{M}_R$ = the category of finitely generated R-modules.

Note that we have $\mathcal{P}_X \subset \mathcal{M}_X$ and $\mathcal{P}_R \subset \mathcal{M}_R$. If the scheme X (resp. ring R) is *regular* (c.f. [Ha]) then every coherent $\mathcal{O}_X$-module (resp. $f.g.$ R-module) has a finite resolution by locally free $\mathcal{O}_X$-modules (resp. projective R-modules) and the inclusions induce isomorphisms on K-theory. We write:

$$K_i(X) = K_i(\mathcal{P}_X), \quad K_i'(X) = K_i(\mathcal{M}_X)$$
$$K_i(R) = K_i(\mathcal{P}_R), \quad K_i'(R) = K_i(\mathcal{M}_R) \ .$$

The tensor product of modules induces biexact functors between these categories making $K_*(X)$ (resp. $K_*(R)$) a graded-commutative ring and $K_*'(X)$ (resp. $K_*'(R)$) a graded module over $K_*(X)$ (resp. $K_*(R)$).

If $f : X \longrightarrow Y$ is a morphism of schemes, or varieties, maps between the K and K' theory of X and Y, both covariant and contravariant, can be constructed, depending on the nature of f. We list some of these below.

1. Arbitrary f. Pull back of vector bundles defines an exact functor $f^* : \mathcal{P}_Y \longrightarrow \mathcal{P}_X$, inducing homomorphisms

$$f^* : K_*(Y) \longrightarrow K_*(X) \ .$$

One can show that K_* is then a contravariant functor taking values in the category of graded-commutative rings.

2. Flat f. Pull back of coherent sheaves is an exact functor, hence there is a map $f^* : K_*'(Y) \longrightarrow K_*'(X)$.

3. Proper f. If f is a projective map, i.e. it factors as $f = p \circ i$ with $i : X \longrightarrow \mathbb{P}_Y^n$ a closed immersion and $p : \mathbb{P}_Y^n \longrightarrow Y$ the projection, then if $\mathcal{A}_X \subset \mathcal{M}_X$ is the subcategory consisting of coherent sheaves $\mathcal{F}$ of $\mathcal{O}_X$-modules such that

$R^i f_* \mathcal{F} = 0$ for $i > 0$, then $f_* : \mathcal{A}_X \longrightarrow \mathcal{M}_Y$ is exact, and the inclusion $\mathcal{A}_X \longrightarrow$ $\mathcal{M}_X$ induces an isomorphism on K-theory. Hence we get

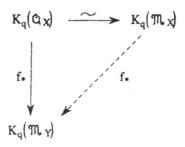

and a well defined map $K'_*(X) \longrightarrow K'_*(Y)$. See [Q] for details. The extension to general proper f may be found in [Gi2], [T2].

K'_* can be viewed as homology with locally compact supports, and K_* as cohomology. The products above are analogous to the cup and cap products in a generalized (co-)homology theory. In particular, the projection formula is true: if $f : X \longrightarrow Y$ is proper, $\alpha \in K_p(Y)$ and $\beta \in K'_q(X)$, then we have that $f_*\big(f^*(\alpha) \cdot \beta\big) = \alpha \cdot f_*(\beta)$.

The analog of Poincaré duality is a consequence of the fact that on a regular scheme X, any coherent sheaf $\mathcal{F}$ of $\mathcal{O}_X$-modules has a finite global resolution by locally free sheaves:

$$0 \longrightarrow \mathcal{P}_n \longrightarrow \mathcal{P}_{n-1} \longrightarrow \ldots \longrightarrow \mathcal{P}_1 \longrightarrow \mathcal{P}_0 \longrightarrow \mathcal{F} \longrightarrow 0 \ .$$

Hence by Quillen's resolution theorem, the inclusion $\mathcal{P}_X \longrightarrow \mathcal{M}_X$ induces an isomorphism $K_*(X) \longrightarrow K'_*(X)$ on K-theory.

Finally the analog for algebraic K-theory of homotopy invariance for a generalized cohomology theory in the topological category is that if $p : \mathbf{A}^1_X \longrightarrow X$ is the projection, then the flat pull-back map $p^* : K'_*(X) \longrightarrow K'_*(\mathbf{A}^1_X)$ is an isomorphism.

4.4. The +-Construction

Another construction, also due to Quillen [Q], of higher K-theory is the *plus construction*. This construction is for the K-theory of rings not general schemes. It should also be noted that the +-construction was the 'first' definition of higher algebraic K-theory.

Let R be a ring (with unit). For $n \geq 1$ embed $GL_n(R) \subset GL_{n+1}(R)$ via

$$A \longrightarrow \left(\begin{array}{c|c} A & 0 \\ \hline 0 & 1 \end{array} \right).$$ Let $GL(R) = \bigcup_{n \geq 1} GL_n(R)$. We can form

(see 4.1) the classifying space $BGL(R)$; recall that $\pi_1\big(BGL(R)\big) = GL(R)$ and $\pi_i\big(BGL(R)\big) = 0$ for $i > 0$. One can attach 2 and 3 cells to $BGL(R)$ forming a new space $BGL(R)^+$ in such a way that

1) $\pi_1 BGL(R)^+ = H_1 BGL(R) = GL(R)^{ab}$.
2) The inclusion $BGL(R) \longrightarrow BGL(R)^+$ is a homology equivalence.
3) The inclusion is universal for maps from $BGL(R)^+$ to H-spaces.

Quillen then defined $K_i(R) = \pi_i BGL(R)^+$ for $i > 0$. It may be shown ([Gr1]) that this definition gives the same groups as the Q-construction applied to the category of finitely generated projective R-modules.

4.5. Examples and Computations

1. *If R is a local ring $K_1(R) = R^*$.* In general, for commutative rings, $K_1(R) = R^* \oplus SK_1(R)$. The splitting is induced by the maps:

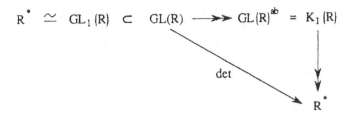

2. *If k is a field $K_2(k) = k^* \otimes_{\mathbf{Z}} k^* / St$ where St is the group of Steinberg relations,* see Lluis-Puebla's lecture III.3, generated by the following elements (we write the group $k^* \otimes k^*$ *multiplicatively*):

$\quad$ a) $x \otimes (1 - x) \qquad x \in k^* - \{1\}$
$\quad$ b) $x \otimes (-x) \qquad x \in k^*$
$\quad$ c) $(x \otimes y)(y \otimes x) \qquad x, y \in k^*$.

The image of $x \otimes y$ in the quotient group is written $\{x, y\}$ and is called a Steinberg symbol.

The map $K_1(k) \otimes K_1(k) = k^* \otimes k^* \longrightarrow K_2$, is none other than the product in K-theory.

3. *Milnor K-theory.* In general one can consider the graded ring which is the quotient of the tensor algebra $\oplus_{p \geq 0} (k^*)^{\otimes p}$ by the relations in 2. above. This is called Milnor K-theory $K_*^M(k)$. One can show [Su] that up to torsion the Milnor K-theory $K_p^M(k)$ of a field k is just the image in $K_p(k)$ of the product $K_1(k)^{\otimes p} \longrightarrow K_p(k)$. However the Milnor K-theory of a field is generally much smaller than the Quillen K-theory. For example if k is a finite extension of $\mathbb{Q}$, $K_p^M(k)$ is torsion for $p > 1$, while for $p \equiv 3(4)$ $K_p(k)$ has non trivial torsion free component.

See the Lectures of Lluis-Puebla, and the book [Mi], as well as [B-T] for further details.

4.6. Connections with Geometry and Arithmetic

There are many connections, both conjectured and proved (but mostly the former), between algebraic K-theory, geometry and arithmetic.

Among the connections with arithmetic, there are the conjectures of Coates and Lichtenbaum ([Co1], [Co2], [L], Bloch and Beilinson [Bei]). See Soulé's article for further discussion of some of these connections.

Here I shall mention one connection with geometry: *Bloch's formula*. Since K_* is a contravariant functor, given a scheme X, one can sheafify the presheaf $U \longrightarrow K_p(U)$ for $p \geq 0$, to obtain a sheaf which we shall denote $K_p(\mathcal{O}_X)$. The stalk of $K_p(\mathcal{O}_X)$ at $x \in X$ may be shown to be $K_p(\mathcal{O}_{X,x})$. Recall that if R is a local ring, $K_1(R) \simeq R^*$, hence $K_1(\mathcal{O}_X) \simeq \mathcal{O}_X^*$, and $H^1(X, \mathcal{O}_X^*) \simeq \mathrm{Pic}(X)$. Generalizing this, we have a theorem of Bloch ([Bl] $p = 2$) and Quillen (all $p \geq 0$):

Theorem "Bloch's Formula". *Let X be a smooth variety over a field. Then there is a natural isomorphism, for all $p \geq 0$,*

$$H^p(X, K_p(\mathcal{O}_X)) \longrightarrow CH^p(X)$$

between the cohomology of X with coefficients in the higher K-theory sheaf $K_p(\mathcal{O}_X)$ and the Chow group of codimension p-cycles modulo rational equivalence.

Proof. [Q]. §7.   □

The product structure induced on the K-theory of locally free sheaves on X by the tensor product gives a product on the graded sheaf $\oplus_{p \geq 0} K_p(\mathcal{O}_X)$.

It follows that $\oplus_{p \geq 0} H^p(X, K_p(\mathcal{O}_X))$ is a commutative ring. On the other hand, if X is a smooth quasi-projective variety, the Chow groups have a geometrically defined product structure. It is proved in [Gr1], (see also [Gi3]) that these are compatible, with a possible sign change depending on the choice of isomorphism.

5. The Map From K_0^{alg} to K_{top}^0

In this chapter we shall construct and study some of the properties of a natural transformation $\eta^0 : K_0^{\mathrm{alg}} \longrightarrow K_{\mathrm{top}}^0$ between functors on the category of algebraic varieties over $\mathbb{C}$.

5.1. Algebraic vs. Topological Bundles

If X is an algebraic variety over $\mathbb{C}$ (i.e. a scheme which is reduced, irreducible, and of finite type over $\mathbb{C}$), we shall write $X(\mathbb{C})$ for the set of $\mathbb{C}$-valued points of X, endowed with the analytic or classical topology; i.e., the topology in which, if $U \subset X$ is an affine open set, isomorphic to a Zariski closed subset of A^n for some n, $U(\mathbb{C}) \subset X(\mathbb{C})$ is an open set with topology induced from the metric topology on $U(\mathbb{C}) \subset A^n(\mathbb{C}) = \mathbb{C}^n$. Note that the classical topology is finer than the Zariski topology. It is straightforward to prove that if $f : X \longrightarrow Y$ is a morphism between varieties, that f induces a continuous map $X(\mathbb{C}) \longrightarrow Y(\mathbb{C})$, so that $X \longrightarrow X(\mathbb{C})$ is a functor from the category of varieties to the category of topological spaces. One may show ([Hi]) that $X(\mathbb{C})$ has a triaaaaaangulation and is therefore a CW complex. Hence given any functor from the category of topological spaces, or CW complexes, such as K^*_{top} or H^*, composing with the functor of $\mathbb{C}$-valued points defines a functor from the category of varieties over $\mathbb{C}$. To simplify notation, we shall write $K^*_{\text{top}}(X)$, $H^*(X, \mathbb{Z})$, etc. for the resulting functors, in place of $K^*_{\text{top}}\big(X(\mathbb{C})\big)$, $H^*\big(X(\mathbb{C}), \mathbb{Z}\big)$, etc.

If $\pi : E \longrightarrow X$ is a geometric vector bundle over a variety X, then $\pi(\mathbb{C}) : E(\mathbb{C}) \longrightarrow X(\mathbb{C})$ is a topological vector bundle. Note that a topological vector bundle $p : V \longrightarrow X(\mathbb{C})$ is isomorphic to $\pi(\mathbb{C}) : E(\mathbb{C}) \longrightarrow X(\mathbb{C})$ for some algebraic bundle if and only if V has a local trivialization $(U_\alpha(\mathbb{C}), \theta_\alpha)$ with $U_\alpha(\mathbb{C})$ the set of $\mathbb{C}$-valued points of a Zariski open $U_\alpha \subset X$ for all α, and such that the transition functions $\theta_{\alpha,\beta} : U_\alpha \cap U_\beta(\mathbb{C}) \longrightarrow GL_n(\mathbb{C})$ are not just continuous, but are also algebraic i.e. the entries of the matrix $\theta_{\alpha\beta}$ are the $\mathbb{C}$-valued functions associated to elements of $\mathcal{O}_X(U_\alpha \cap U_\beta)$.

The following theorem, the proof of which is straightforward, describes the properties of the operation $E \longrightarrow E(\mathbb{C})$.

Theorem . 1. *Let X be an algebraic variety over $\mathbb{C}$. Then $E \longrightarrow E(\mathbb{C})$ defines an exact functor T_X from the category of algebraic vector bundles over X to the category of topological vector bundles over $X(\mathbb{C})$.*

2. If E and F are algebraic bundles over a variety X, then $(E \otimes F)(\mathbb{C}) \simeq E(\mathbb{C}) \otimes F(\mathbb{C})$.

3. If $f : X \longrightarrow Y$ is a morphism of varieties over $\mathbb{C}$, and

$$f^*_{\text{alg}} : \text{Vect}(Y) \longrightarrow \text{Vect}(X)$$

$$f^*_{\text{top}} : \text{TopVect}(Y(\mathbb{C})) \longrightarrow \text{TopVect}(X(\mathbb{C}))$$

are the (exact) pull back functors on the categories of algebraic and topological bundles respectively, there is a natural isomorphism of functors.

$$T_X \cdot f^*_{\text{alg}} \simeq f^*_{\text{top}} \cdot T_Y$$

5.2. Construction of η^0

Theorem . *The functors T_X, for X in the category of varieties over $\mathbb{C}$, induce a natural transformation :*

$$\eta^0 : K_0(T_-) : K_0 \longrightarrow K^0_{\text{top}}$$

between contravariant commutative ring valued functors on the category of varieties over $\mathbb{C}$.

Proof. By parts 1 and 2 of theorem 5.1, for each X there is a homomorphism of rings :

$$\eta^0_X = K_0(T_X) : K_0(X) \longrightarrow K^0_{\text{top}} .$$

By part 2, if $f : X \longrightarrow Y$ is a morphism of varieties,

$$K^0_{\text{top}}(f) \cdot \eta^0_Y = \eta^0_X \cdot K^{\text{alg}}_0(f) .$$

Remark. There are other ways of defining η^0.
1) There is a continuous map of topological spaces $\mu : X(\mathbb{C}) \longrightarrow X$, and a map $\mu^* : \mathcal{O}_X \longrightarrow \mathcal{C}_X$ (where $\mathcal{C}_X$ is the sheaf of continuous complex valued functions on $X(\mathbb{C})$) induced by the inclusion $\mathcal{O}_X(U) \subset \mathcal{C}_X(U(\mathbb{C}))$ for each affine open subset $U \subset X$. Given a geometric vector bundle E over X, and $\mathcal{E}$ its sheaf of (algebraic) sections, then $\mathcal{E} \otimes_{\mathcal{O}_X} \mathcal{C}(X)$ is a locally free sheaf of $\mathcal{C}_X$-modules on $X(\mathbb{C})$. It may be checked that $\mathcal{E} \otimes_{\mathcal{O}_X} \mathcal{C}(X)$ is naturally isomorphic to the sheaf of continuous sections of $E(\mathbb{C})$.

2) If $X = \text{Spec}(A)$ is affine, and E, $\mathcal{E}$ are as above, then $\mathcal{E}(X)$ is a finitely generated projective A-module. Hence $\mathcal{E}(X) \otimes_A \mathcal{C}(X)$ is a $f.g.$ projective $\mathcal{C}(X) = \mathcal{C}(X(\mathbb{C}))$-module, which determines a vector bundle by (2.4).

5.3

As we saw in the previous section, η^0 is a natural transformation of contravariant functors, preserving products. As discussed previously, both K^{alg}_0 and K^0_{top} may also be viewed as *covariant* functors on the category of all non-singular *projective* varieties over $\mathbb{C}$. A version of the Grothendieck-Riemann-Roch theorem ([B5]) ([B-F-M-2]) says that η^0 is also a natural transformation of *covariant* functors. In particular, if X is a smooth projective variety, applying this result to the canonical map $\pi : X \longrightarrow \text{Spec}(\mathbb{C})$, we find that if E is a vector bundle over X.

$$\mathcal{X}(X, E) = \sum_{i \geq 0} (-1)^i \dim_{\mathbb{C}} H^i(X, E)$$
$$= \pi_*([E]) \in K_0(\mathbb{C}) \simeq \mathbb{Z}$$

depends only on $\eta^0([E])$, i.e. the class of E viewed as a topological bundle. Since the Chern character

$$\text{ch} : K^0_{\text{top}}(X) \longrightarrow H^*_{\text{top}}(X, \mathbb{C})$$

is an isomorphism, this means that $\mathcal{X}(X, E)$ is determined by

$$\text{ch}(E) \in H^*_{\text{top}}(X, Q) .$$

Specifically we have:

Theorem . *(Hirzebruch, [Hi]). Let X be a smooth projective variety over $\mathbb{C}$ of dimension n. Let E be an algebraic bundle over X. Then*

$$\mathcal{X}(X, E) = \big[\text{ch}(E) \cup \text{Td}(X)\big]_{2n}$$

where $[\ ]_{2n}$ means the component in $H^{2n}(X, \mathbb{Q}) \simeq \mathbb{Q}$ and $\text{Td}(X) \in H^*(X, \mathbb{Q})$ is the Todd class of X, a polynomial in the Chern classes of the tangent bundle of X.

In particular if X is a smooth projective curve over $\mathbb{C}$ over genus g, then

$$\mathcal{X}(X, E) = r(1 - g) + d$$

where $r = \text{rank }(E)$, and $d = \text{degree }(E)$ (i.e. of the top exterior power $\Lambda^r E$).

Though the Riemann-Roch theorem shows that η^0 gives us significant information about the algebraic K-theory of a variety, we shall see that this information is far from complete. In particular η^0 is neither injective nor surjective in general.

5.4. Basic Ideas of Hodge Theory

Before continuing our discussion of the injectivity and surjectivity of η^0, we need to review, briefly, some of the ideas of Hodge theory. For details, see [GH], [De1], [De2].

Let X be a non singular projective variety over $\mathbb{C}$. Then $X(\mathbb{C})$ is a compact complex manifold, and the space of C^∞ differential k-forms $A^k(X) = A^k(X(\mathbb{C}))$ has a decomposition:

$$A^k(X) = \bigoplus_{p+q=k} A^{p,q}(X)$$

with $\alpha \in A^{p,q}(X)$ if and only if in each coordinate chart with local holomorphic coordinates $z_1, \ldots, z_n$, α can be written as a sum of forms of the form $f dz_{i_1} \wedge \ldots \wedge dz_{i_p} \wedge d\bar{z}_{j_1} \wedge \ldots \wedge d\bar{z}_{j_q}$. Write

$$F^p A^k(X) = \bigoplus_{\substack{i \geq p \\ i+j=p}} A^{i,j}(X) .$$

Then $F^p A^*(X)$ is a subcomplex of $A^*(X)$, i.e. if $\alpha \in F^p A^k$ then $d\alpha \in F^p A^{k+1}$. The quotient complexes $\text{Gr}^p A^*(X)$ are isomorphic to the Dolbeault complexes $A^{p,0} \xrightarrow{\bar{\delta}} A^{p,1}(X) \xrightarrow{\bar{\delta}} \ldots$. Hence $H^q(\text{Gr}^p A^*(X)) \simeq H^q_{\bar{\delta}}(X, \Omega^p)$, where Ω^p is the p-th exterior power of the cotangent bundle. A key result is that the natural map $H^*(F^p A^*(X)) \longrightarrow H^*(A^*(X))$ is injective, and hence that there is a (non-canonical) isomorphism, for $k \geq 0$,

$$H^k(X, \mathbb{C}) \simeq \bigoplus_{p+q=k} H^{p,q}(X, \mathbb{C})$$

with $H^{p,q}(X, \mathbb{C}) \simeq H^q(X, \Omega^p)$. This direct sum decomposition is called the Hodge decomposition. If we write

$$F^p H^k(X, \mathbb{C}) = H^k(F^p A^*(X)) = \bigoplus_{\substack{i \geq p \\ i+j=k}} H^{i,j}(X, \mathbb{C})$$

(F^0 is called the *Hodge filtration*). Then, for $k = p + q$,

$$H^{p,q}(X, \mathbb{C}) = F^p H^k(X, \mathbb{C}) \cap \overline{F^q H^k(X, \mathbb{C})}$$

(here "—" denotes complex conjugation). Since the cup product on cohomology is induced, via the deRham isomorphism, by wedge product of forms, one sees easily that $F^p \cup F^q \subset F^{p+q}$; (note we omitted the $H^*(X, \mathbb{C})$).

If $f : X \longrightarrow Y$ is a map between non-singular projective varieties, and $f^* : H^*(Y, \mathbb{C}) \longrightarrow H^*(X, \mathbb{C})$ the induced map on cohomology, then

$$f^*(F^p H^*(Y, \mathbb{C})) \subset F^p H^*(X, \mathbb{C}) \ .$$

For us one of the important features of the Hodge filtration is:

Theorem . *Let X be a non-singular projective variety over $\mathbb{C}$. If E is an algebraic vector bundle over X, then*

$$c_p(E) \in F^p H^{2p}(X, \mathbb{C}) \cap H^{2p}(X, \mathbb{R}) \subset H^{2p}(X, \mathbb{C}) \ .$$

See [G-H] 3.3 for a proof. Note that the theorem is true more generally for any holomorphic bundle over a compact complex manifold.

5.5. Hodge Theory and η^0

It follows from theorem 5.4 that the image of $\eta^0 : K_0(X) \longrightarrow K^0_{\text{top}}(X)$, for X a non-singular projective variety over $\mathbb{C}$, is contained in the subgroup consisting of those elements α such that $c_p(\alpha) \in H^{2p}(X, \mathbb{Z}) \cap F^p H^{2p}(X, \mathbb{C})$ for all $p \geq 0$.

The *Hodge conjecture* states that if X is a non-singular projective variety over $\mathbb{C}$, then $H^{2p}(X, Q) \cap F^p H^{2p}(X, \mathbb{C})$ is spanned by the fundamental classes of algebraic cycles. The topological Chern character induces an isomorphism ch : $K^0_{\text{top}}(X) \otimes_{\mathbb{Z}} \mathbb{Q} \longrightarrow \oplus_{p \geq 0} H^{2p}(X, \mathbb{Q})$, and the algebraic Chern character an isomorphism ([SGA6] exp XIV):

$$\text{ch} : K_0(X) \otimes_{\mathbb{Z}} \mathbb{Q} \simeq \bigoplus_{p \geq 0} CH^p(X) \otimes_{\mathbb{Z}} Q \ .$$

It follows that the Hodge conjecture is equivalent to the assertion that the image of $\eta^0 : K_0(X) \otimes_{\mathbb{Z}} \mathbb{Q} \longrightarrow K_{\text{top}}^0(X) \otimes_{\mathbb{Z}} \mathbb{Q}$ is precisely equal to the subspace of all elements α such that $c_p(\alpha) \in F^p H^{2p}(X, \mathbb{C})$ for all $p \geq 0$.

Note that there are many examples of varieties, for example $K3$ surfaces or the product of two curves for positive genus, for which there is a $p \geq 0$ such that $H^{2p}(X, \mathbb{Q})$ is *not* contained in $F^p H^{2p}(X, \mathbb{C})$, and hence for which η^0 cannot be surjective.

Turning to the question of the injectivity of η^0, we find that for any non-singular projective algebraic curve X over $\mathbb{C}$ of genus $g > 0$,

$$K_0(X) \simeq \mathbb{Z}^2 \oplus \{\mathbb{C}^g/\mathbb{Z}^{2g}\}$$

while $K_{\text{top}}^0(X) \simeq \mathbb{Z}^2$. Hence the kernel of η^0 in this case is $\mathbb{C}^g/\mathbb{Z}^{2g}$, a compact complex torus of dimension $g > 0$. One can view the points of this torus as parametrizing inequivalent holomorphic structures on topological vector (and in particular line) bundles. For a curve, this torus is $H^1(X, \mathcal{O}_X)/H^1(X, \mathbb{Z})$, or equivalently $H^1(X, \mathbb{C})/\{F^1(X, \mathbb{C}) + H^1(X, \mathbb{Z})\}$. This may be seen using the exponential sequence of sheaves of abelian groups on $X(\mathbb{C})$:

$$0 \longrightarrow \mathbb{Z} \xrightarrow{2\pi i} \mathcal{O}_X \xrightarrow{\exp} \mathcal{O}_X^* \longrightarrow 1$$

and the corresponding cohomology sequence

$$\longrightarrow H^1(X, \mathbb{Z}) \longrightarrow H^1(H, \mathcal{O}_X) \longrightarrow H^1(X, \mathcal{O}_X^*) \approx \text{Pic}(X) \xrightarrow{\partial} H^2(X, \mathbb{Z}) \longrightarrow \ .$$

In general the Chern classes for vector bundles on a non-singular projective variety X over $\mathbb{C}$, $c_p : K_0(X) \longrightarrow H^{2p}(X, \mathbb{Z})$ factor through the Deligne cohomology groups $H_{\mathcal{D}}^{2p}(X, \mathbb{Z})$. These groups fit in an exact sequence

$$0 \longrightarrow \frac{H^{2p-1}(X, \mathbb{C})}{F^p H^{2p-1}(X, \mathbb{C}) + H^{2p-1}(X, \mathbb{Z})} \longrightarrow H_{\mathcal{D}}^{2p}(X, \mathbb{Z})$$
$$\longrightarrow H^{2p}(X, \mathbb{Z}) \cap F^p H^{2p}(X, \mathbb{C}) \longrightarrow 0$$

i.e. they are an extension of the subgroup of $H^{2p}(X, \mathbb{Z})$ which the Hodge conjecture predicts consists (up to torsion) of cycle classes of algebraic cycles, by a compact complex torus $J^p(X)$. Hence, for $p \geq 0$, there is a homomorphism

$$\bar{\eta}^0 : \ker(\eta^0) \longrightarrow \bigoplus_{p \geq 0} J^p$$

which can be shown to be non-trivial in general. See [Har].

One can of course ask whether $\bar{\eta}^0$ is injective. Again the answer is no, even for X the product of two curves of positive genus. See [Mu].

6. The Map From Higher Algebraic K-Theory to Topological K-Theory

6.1. The Affine Case

If $X = \mathrm{Spec}(R)$, $R = \mathcal{O}(X)$, is an affine scheme, recall that there is an equivalence between the categories $\mathcal{P}_X$ of vector bundles on X, i.e., locally free sheaves of $\mathcal{O}_X$-modules, and $\mathcal{P}_R$ of projective R-modules. We saw in §4 that for $q \geq 0$,

$$K_q(X) \simeq \pi_{q+1}(BQ\mathcal{P}_X) = \pi_q\left(K_0(R) \times BGL(R)^+\right) \ .$$

Having already constructed the map $K_0(X) \longmapsto K^0_{\mathrm{top}}(X)$, we turn now to the maps $\eta^q : K_q \longrightarrow K^{-q}_{\mathrm{top}}(X)$ for $q > 0$. Here by $K^*_{\mathrm{top}}(X)$ we mean the topological K-theory of the topological space $X(\mathbb{C})$ consisting of the $\mathbb{C}$-valued points of X, given the "classical" topology. The ring R embeds naturally as a subring of $\mathcal{C}(X)$, the ring of all continuous $\mathbb{C}$-valued functions on $X(\mathbb{C})$. Hence there is a homomorphism of groups $GL(R) \longrightarrow GL\big(\mathcal{C}(X)\big)$, and a map of classifying spaces $BGL(R) \longrightarrow BGL\big(\mathcal{C}(X)\big)$. It is important to note here that we give $GL(R)$ the discrete topology, and $GL\big(\mathcal{C}(X)\big)$ the compact open topology under the natural identification of $GL\big(\mathcal{C}(X)\big)$ with the group $GL(\mathbb{C})^X$ of continuous maps from X to $GL(\mathbb{C})$.

Lemma . *There is a natural homotopy equivalence*

$$BU^X \longrightarrow BGL\big(\mathcal{C}(X)\big) \ .$$

Proof. As previously remarked, the evaluation map $X \times GL\big(\mathcal{C}(X)\big) \longrightarrow GL(\mathbb{C})$ identifies $GL\big(\mathcal{C}(X)\big)$ with $GL(\mathbb{C})^X$. Hence $BGL\big(\mathcal{C}(X)\big)$ is homeomorphic to $B\big(GL(\mathbb{C})^X\big)$. The functor on spaces, $Y \longmapsto Y^X$ preserves fibration sequences, and maps contractible spaces to contractible spaces, hence $\big(BGL(\mathbb{C})\big)^X$ is a classiifying space for $GL(\mathbb{C})^X$ and there is a map, unique up to homotopy, $\big(BGL(\mathbb{C})\big)^X \longrightarrow B\big(GL(\mathbb{C})^X\big)$. Since the inclusion $U \hookrightarrow GL(\mathbb{C})$ induces a homotopy equivalence $BU^X \longrightarrow \big(BGL(\mathbb{C})\big)^X$, on composing we find a homotopy equivalence

$$BU^X \longrightarrow BGL\big(\mathcal{C}(X)\big)$$

as desired. $\qquad\qquad\qquad\qquad\qquad\qquad\qquad\qquad\qquad\qquad\qquad\qquad\qquad\qquad$ □

Theorem . *On the category of affine varieties X over $\mathbb{C}$, there is a natural transformation of functors to the homotopy category of spaces:*

$$\eta_X : BGL\big(\mathcal{O}(X)\big)^+ \longrightarrow BU^X \ .$$

Proof. Composing the natural map $BGL\big(\mathcal{O}(X)\big) \longrightarrow BGL\big(\mathcal{C}(X)\big)$ with the homotopy inverse of the map of the lemma, we obtain a natural transformation of homotopy functors $BGL(\mathcal{O}(X)) \longrightarrow BU^X$. Since BU, and hence BU^X, is an H-space, by the universal mapping property of the plus construction, this map factors through a map $\eta : BGL(\mathcal{O}(X))^+ \longrightarrow BU^X$ which is unique up to homotopy, as desired.   □

Definition . For $q > 0$, η^q is the induced map

$$K_q(X) = \pi_q BGL(\mathcal{O}(X)^+) \longrightarrow \pi_q(BU^X) = K_{\text{top}}^{-q}(X) \ .$$

6.2. The Quasi-projective Case

If X is a quasi-projective variety over $\mathbb{C}$, or more generally any quasi-projective scheme over a ring, recall that X admits a Jouanolou covering, i.e. a map $p : T \longrightarrow X$ with T affine, such that there exists an affine open covering $X = \cup U_\alpha$ and isomorphisms $p^{-1}(U_\alpha) \longrightarrow \mathbf{A}_{U_\alpha}^{n_\alpha}$ of varieties over X. By (2.5) the pullback map

$$p_{\text{top}}^* : K_{\text{top}}^{-q}(X) \longrightarrow K_{\text{top}}^{-q}(T)$$

is an isomorphism. We now define

$$\eta_X^q = p_{\text{top}}^{*\,-1} \eta_T^q p^* \ .$$

Of course this definition depends, a priori, on the choice of Jouanolou cover.

Theorem . *(i) η^q is independent of the choice of $p : T \longrightarrow X$.*
(ii) η^* is a natural transformation of functors from the category of quasi-projective varieties over $\mathbb{C}$ to the category of graded abelian groups .

Proof. If $p : T \longrightarrow X$ and $p' : T' \longrightarrow X'$ are two Jouanolou covers, then we can form their fibre product:

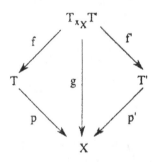

which is again a Jouanolou cover of X. Since f has fibres which are affine spaces, $f^* : K_{\text{top}}^*(T) \longrightarrow K_{\text{top}}^*(T \times_X T')$ is an isomorphism. Hence, by theorem 6.1,

$$p_{\text{top}}^{*\,-1} \eta_T^* p^* = p_{\text{top}}^{*\,-1} f_{\text{top}}^{*\,-1} \eta_{T \times T'} f^* p^*$$

$$= g_{\text{top}}^{*\,-1} \eta_{T \times T'} g^*$$

and we are done, by symmetry.

Turning to part ii), if $f : X \longrightarrow Y$ is a map of varieties, and $p : T \longrightarrow Y$ is a Jouanolou cover, then the fibre product $X \times_Y T$ will not in general be a Jouanolou cover of X. However, since p is an affine map (i.e. the inverse image of an affine open is affine), if $q : S \longrightarrow X$ is a Jouanolou cover, then $S : S \times_X X \times_Y T = S \times_Y T$ will be a Jouanolou cover of X. Hence we may suppose, replacing S by S', that we have a commutative diagram

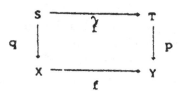

with p and q Jouanolou coverings. Part ii) is now proved by a diagram chase, using theorem 6.1 and the homotopy invariance of $K^*_{\text{top}}(X)$

Remark. (i) Thomason has shown that a Jouanolou torsor exists not just for a quasi-projective scheme, but for any scheme having an ample family of line bundles ; this includes all, possibly non quasi-projective, regular schemes. The methods above extend to define the natural transformation η for all such varieties over $\mathbb{C}$.

(ii) There are at least two other methods of constructing η which work for general varieties over $\mathbb{C}$. The first of these was described in [Gi1]. The second method, which has not been described in detail in the literature consist of viewing the category of topological vector bundles on a space X as a topological category $\mathcal{P}^{\text{top}}(X)$, i.e. a category in which the hom-sets have a topology. One forms the Q-construction, this time as a topological category, and shows that $\Omega BQ\mathcal{P}^{\text{top}}(X) \simeq (\mathbb{Z} \times BU)^X$. Then $\mathcal{P}(X) \longrightarrow \mathcal{P}^{\text{top}}(X)$ is a continuous exact functor, hence induces a map

$$K_p(X) = \pi_{p+1}\big(BQ\mathcal{P}(X)\big) \longrightarrow \pi_{p+1}\big(BQ\mathcal{P}^{\text{top}}(X)\big)$$
$$= \pi_p\big((\mathbb{Z} \times BU)^X\big) = K^{-p}_{\text{top}}(X) \ .$$

$\square$

A description of this type of construction is contained in the U.I.C. thesis of M. Paluch.

6.3. The Image of η

Theorem . *If X is a smooth projective variety over $\mathbb{C}$, then for $q > 0$, the map*

$$\eta^q_X : K_q(X) \longrightarrow K^{-q}_{\text{top}}(X)$$

has torsion image.

Proof. One may show ([G-A] for example) that the Chern character

$$\text{ch} : K_{\text{top}}^{-q}(X) \otimes_{\mathbb{Z}} \mathbb{Q} \longrightarrow \bigoplus_i H^{2i-q}(X, \mathbb{Q})$$

is an isomorphism. Therefore it suffices to show that the composition $\text{ch} \circ \eta^q$ is zero. Let $p : T = \text{Spec}(R) \longrightarrow X$ be a Jouanolou covering. Since $p^* : H^*(X, \mathbb{Z}) \longrightarrow H^*(T, \mathbb{Z})$ is an isomorphism, it suffices to show that, for all $\beta \in K_q(T)$, $\text{ch}_i(\eta_T^q(\beta)) \in H^{2i-q}(T, Q)$ is zero for $i \geq q/2$.

Recall that η_T^* is induced by factoring the natural map $BGL(R) \longrightarrow BGL(\mathbb{C})^{T(\mathbb{C})}$ through $BGL(R)^+$. By the adjunction property of mapping spaces this factorization corresponds to a factorization, up to homotopy, of the evaluation map $\mu : T(\mathbb{C}) \times BGL(R) \longrightarrow BGL(\mathbb{C})$ through a map

$$\rho : (T(\mathbb{C}) \times BGL(R)^+)/(T(\mathbb{C}) \times \{*\}) \longrightarrow BGL(\mathbb{C}).$$

The element $\beta \in K_q(T) = \pi_q(BGL(R)^+)$ therefore induces a homotopy class of maps:

$$[\beta] : T(\mathbb{C}) \times S^q/T(\mathbb{C}) \times \{*\} = \Sigma^q T(\mathbb{C}) \longrightarrow BGL(\mathbb{C})$$

and then

$$\text{ch}_i(\eta_T^q(\beta)) = [\beta]^*(\text{ch}_i) \in H^{2i}(\Sigma^q T(\mathbb{C}), \mathbb{Q}) = H^{2i-q}(T, \mathbb{Q}) ,$$

where $\text{ch}_i \in H^{2i}(BGL(\mathbb{C}), \mathbb{Q}) = H^{2i}(BU, \mathbb{Q})$ is the universal i-th Chern character. We can reformulate this construction as follows. The Hurewicz homomorphism associates to β an element $h^+(\beta) \in H_q(BGL(R)^+, \mathbb{Z})$, while $\rho^*(\text{ch}_i) \in H^{2i}(T(\mathbb{C}) \times BGL(R)^+, \mathbb{Q})$. The slant product ([Sp] p 287) $\rho^*(\text{ch}_i)/h^+(\beta)$ is equal to $\text{ch}_i(\beta) \in H^{2i-q}(T, \mathbb{Q})$, since the isomorphism $H^{2i}(\Sigma^q T(\mathbb{C}), Q) \longrightarrow H^{2i-q}(T, \mathbb{Q})$ is given by slant product with the generator of $H_q(S^q, \mathbb{Z})$ which is the image under the Hurewicz homomorphism of the canonical generator of $\pi_q(S^q)$. Now by construction $H_*(BGL(R), \mathbb{Z}) \xrightarrow{\sim} H_*(BGL(R)^+, \mathbb{Z})$, so $h^+(\beta)$ is the image of a class $h(\beta) \in H_q(BGL(R), \mathbb{Z})$ while $\text{ch}_i \in H^{2i}(BGL(\mathbb{C}), \mathbb{Q})$ pulls back to a class $\sigma^*(\text{ch}_i) \in H^{2i}(T(\mathbb{C}) \times BGL(R), \mathbb{Q})$. By a diagram chase, we see that $\text{ch}_i(\beta) = \sigma^*(\text{ch}_i)/h(\beta)$.

The evaluation maps, for $n \geq 1$, $T \times GL_n(R) \longrightarrow GL_n$ induce maps of simplicial schemes

$$T \times N.GL_n(R) \longrightarrow N.GL_n .$$

Taking complex points, and even geometric realizations, give us the maps

$$T(\mathbb{C}) \times BGL_n(R) \longrightarrow BGL_n(\mathbb{C}) .$$

As remarked previously we can compute the cohomology of these topological spaces as the cohomology of sheaves on the corresponding simplicial schemes. Since $GL(\mathbb{C}) = \bigcup_n GL_n(\mathbb{C})$, for some $n \geq 1$, $h(\beta)$ lifts to a class $h_n(\beta) \in H_q(BGL_n(R), \mathbb{Z})$, while the i-th universal Chern class restricts to a class $\sigma_n^*(\text{ch}_i) \in H^{2i}(N.GL_n, \mathbb{Q})$. In [De2], Deligne shows that $H^*(N.GL_n, \mathbb{Z})$ is a polynomial ring on $c_i \in H^{2i}(N.GL_n, \mathbb{Z})$ for $1 \leq i \leq n$, and that $c_i \in F^i H^{2i}(N.GL_n, \mathbb{C})$. Hence

$$\sigma_n^*(\mathrm{ch}_i) \in F^i H^{2i}\big(T \times N.GL_n(R), \mathbb{C}\big) \cap H^{2i}\big(T \times N.GL_n(R), \mathbb{Q}\big) \ .$$

Since the simplicial scheme $T \times N.GL_n(R)$ is the product of the scheme T with a discrete simplicial set, both its integral cohomology, and the i-th stage of the Hodge filtration on complex cohomology are computed by cochains on $N.GL_n(R)$ with coefficients in the complexes computing either of these cohomologies for T. Hence

$$\sigma^*(\mathrm{ch}_i)/h(\beta) = \sigma_n^*(\mathrm{ch}_i)/h_n(\beta) \in H^{2i-q}(Z, \mathbb{Q}) \cap F^i H^{2i-q}(T, \mathbb{C}) \ .$$

However the map $p^* : H^*(X, \mathbb{Z}) \longrightarrow H^*(T, \mathbb{Z})$ has fibres which are affine spaces, hence induces an isomorphism on cohomology and by ( ) preserves the Hodge filtration strictly. Therefore

$$H^{2i-q}(T, \mathbb{Q}) \cap F^i H^{2i-q}(T, \mathbb{C}) \simeq H^{2i-q}(X, \mathbb{Q}) \cap F^i H^{2i-q}(X, \mathbb{C})$$
$$\subset H^{2i-q}(X, \mathbb{R}) \cap \bigoplus_{\substack{j+k=2i-q \\ j \ge i}} H^{j,k}(X, \mathbb{C}) = \{0\} \ .$$

Hence $\mathrm{ch}_i(\beta) = 0$.

7.

7.1. K-Theory with Coefficients

We saw in chapter six that for smooth projective varieties over $\mathbb{C}$, the map η^q from $K_q(X)$ to $K_{\mathrm{top}}^{-q}(X)$ has torsion image for $q > 0$. This suggests that in order to compute, or at least obtain information about, the K-theory of algebraic varieties, it will be necessary to examine both the torsion in K-theory, and the kernels of the maps η^q. The kernel of η has been investigated using regulator maps, which are outside the scope of these notes; see [Bei], [R-S-S], [Ka2]. In this last section we shall consider the torsion in algebraic K-theory. To do this it will be useful to introduce K-theory with coefficients.

If $\mathbf{E}$ is an exact category, and $n \in \mathbb{N}$, there is an exact functor

$$\mu_n : \mathbf{E} \longrightarrow \mathbf{E}$$
$$P \longmapsto P^{\oplus n}$$

and hence a map $BQ\mu_n : BQ\mathbf{E} \longrightarrow BQ\mathbf{E}$.

Definition . If F is the homotopy fibre of $BQ\mu_n$, we set $K_p(\mathbf{E}, \mathbb{Z}/n) = \pi_p(F)$.

Clearly $K_*(\mathbf{E}, \mathbb{Z}/n)$ is covariant with respect to exact functors between exact categories.

We now define $K_p(R, \mathbb{Z}/n)$ for a ring R or $K_p(X, \mathbb{Z}/n)$ for a scheme X as the K-theory with coefficients of the corresponding scheme. Similarly one defines $K_{\mathrm{top}}^*(X, \mathbb{Z}/n)$ for a topological space X, and the natural transformation η induces a natural transformation between functors on the category of varieties over $\mathbb{C}$:

$$\eta^q : K_q(X, \mathbb{Z}/n) \longrightarrow K_{\text{top}}^{-q}(X, \mathbb{Z}/n) .$$

Again it is natural to ask how close η^q is to being an isomorphism.

For $q = 0$, we are again essentially trying to compute the image of the cycle class map to $H^*(X, \mathbb{Z}/n)$, which is governed by the Hodge and Tate conjectures; in particular we know that η^0 cannot be an isomorphism for general projective varieties.

Let us start our discussion of η^q for $q \neq 0$ with the case of finite fields. The following result was one of the starting points for algebraic K-theory.

Since $BQ\mathbf{E}$ is an H-space with addition operation induced by the direct sum functor $\oplus : \mathbf{E} \times \mathbf{E}$, $BQ\mu_n$ induces multiplication by n on $\pi_* BQ\mathbf{E}$. Hence the long exact homotopy sequence for the map $BQ\mu_n$ is:

$$\ldots \longrightarrow K_p(\mathbf{E}) \xrightarrow{n} K_p(\mathbf{E}) \longrightarrow K_p(\mathbf{E}, \mathbb{Z}/n) \longrightarrow K_{p-1}(\mathbf{E}, \mathbb{Z}/n) \xrightarrow{n} \ldots .$$

Therefore there are short exact sequences

$$0 \longrightarrow K_p(\mathbf{E}) \otimes \mathbb{Z}/n \longrightarrow K_p(\mathbf{E}, \mathbb{Z}/n) \longrightarrow n - \text{Torsion}(K_{p-1}(\mathbf{E})) \longrightarrow 0 .$$

It follows immediately that $K_p(\mathbf{E}, \mathbb{Z}/n)$ is killed by multiplication by n^2. (One can show that with a few possible exceptions, $K_*(E, \mathbb{Z}/n)$ is killed by n; we shall not use this fact however.)

There is an alternate approach to the definition of $K_*(\mathbf{E}, \mathbb{Z}/n)$ in which $\pi_p(X, \mathbb{Z}/n)$, for general spaces X and $p \geq 2$, is defined as the set of homotopy classes of maps from the Moore space $M^p(n)$ to X. Here $M^p(n)$ is the mapping cone of the map $S^{p-1} \longrightarrow S^{p-1}$ inducing multiplication by n on π_p. See [Bro], [N].

Theorem . *(Quillen, [Q1]). Let $\mathbb{F}_q$ be the finite field with $q = p^n$ elements for p prime. Then*

$$K_i(\mathbb{F}_q) = \begin{cases} \mathbb{Z} & i = 0 \\ 0 & i > 0 \quad\quad even \\ \mathbb{Z}/(q^k - 1) & i = 2k - 1 \quad k \geq 1 \end{cases}$$

and

$$K_i(\bar{\mathbb{F}}_q) = \begin{cases} \mathbb{Z} & i = 0 \\ 0 & i > 0 \quad\quad even \\ \mathbb{Q}/\mathbb{Z}[1/p] & i = 2k - 1 \quad k \geq 1 \end{cases} .$$

Using the long exact sequence for K-theory with coefficients one obtains

Corollary. If $(n, p) = 1$,

$$K_i(\bar{\mathbb{F}}_p, \mathbb{Z}/n) = \begin{cases} \mathbb{Z}/n & i \text{ even} \\ 0 & i \text{ odd} \end{cases}$$

The striking feature of this corollary is that $K_*(\overline{\mathbb{F}}_p, \mathbb{Z}/n)$ looks like the mod n topological K-theory of a point. This observation, together with some partial computations of the K-theory of rings of integers in number fields led to the conjectures of Lichtenbaum and Quillen relating K-theory with coefficients to etale cohomology. For details see [Q3], [L]. In order to avoid a discussion of etale cohomology or of etale topological K-theory (see [Fr]), here we shall discuss two particular cases :

Algebraically closed fields:

If F is an algebraically closed field, and $1/n \in F$, then

$$K_i(F, \mathbb{Z}/n) = \begin{cases} \mathbb{Z}/n & i \text{ even} \\ 0 & i \text{ odd} \end{cases}$$

Varieties over $\mathbb{C}$:

If X is a nonsingular variety over the complex numbers then the map $\eta^q :$ $K_q(X, \mathbb{Z}/n) \to K_{\text{top}}^{-q}(X, \mathbb{Z}/n)$ is an isomorphism if q is sufficiently large; perhaps if $q > 2\dim(X)$.

In the case of algebraically closed fields, the conjecture has been proved by Suslin, as we shall see in §7.2–7.4.

For varieties over $\mathbb{C}$, or more generally regular schemes, the conjecture is still open, though Thomason has made significant progress ([Th]).

We shall finish this section with a brief outline of what Thomason's results tell us about the K-theory of varieties over $\mathbb{C}$.

Let ζ be a primitive n-th root of 1. Since $\{\zeta\} \in K_1(k)$ is n-torsion, there exists $\beta \in K_2(k, \mathbb{Z}/n)$ such that $\delta(\beta) = \zeta$ under $\delta : K_2(k, \mathbb{Z}/n) \longrightarrow n - \text{Torsion}(K_1(k))$, which is in fact unique. Under η^q, β maps to the generator in K_{top}^{-2} (point); i.e., the element which, via the cup product, induces the Bott periodicity isomorphism $K_{\text{top}}^{-q} \longrightarrow K_{\text{top}}^{-q-2}$. Thomason shows that the induced map

$$K_*(X, \mathbb{Z}/n)\left[\beta^{-1}\right] \longrightarrow K_{\text{top}}^*(X, \mathbb{Z}/n)$$

is an isomorphism.

This result may be interpreted as saying that the failure of Bott periodicity for algebraic K-theory is responsible for the failure of η^q to be an isomorphism. Note that this result is close to the conjecture, for if one knew that multiplication by β gave an isomorphism $K_q(X, \mathbb{Z}/n) \longrightarrow K_{q+2}(X, \mathbb{Z}/m)$ for $q > N$ for some N, then it would follow that $K_q(X, \mathbb{Z}/n) \longrightarrow K_{\text{top}}^{-q}(X, \mathbb{Z}/n)$ is an isomorphism for $q > N$.

7.2. Divisor Class Groups

If X is a smooth connected curve over an algebraically closed field k, the *divisor group* $\mathrm{Div}(X)$ of X is the free abelian group on the rational, i.e., non generic, points of X. If $f \in k(X)$ is a rational function on X, its divisor $\mathrm{div}(f)$ is $\sum_{P \in X} n_P[P] \in \mathrm{Div}(X)$, with n_P equal to the order of vanishing of f at the point P. Clearly $\mathrm{div}(fg) = \mathrm{div}(f) + \mathrm{div}(g)$, so that the divisors of functions form a subgroup $\mathrm{Prin}(X) \subset \mathrm{Div}(X)$, the subgroup of principal divisors. If $D = \Sigma n_P[P]$ is a divisor, its *degree* is $\Sigma n_P \in \mathbb{Z}$. If we write $\mathrm{Div}_0(X) \subset \mathrm{Div}(X)$ for the subgroup consisting of divisors of degree zero, it is a classical theorem that $\mathrm{Prin}(X) \subset \mathrm{Div}_0(X)$ when X is projective.

The *divisor class group* $C\ell(X)$ is the quotient $\mathrm{Div}(X)/\mathrm{Prin}(X)$. If X is projective, we also have $C\ell_0(X) \subset C\ell(X)$ the subgroup of divisor classes of degree 0. This construction is directly analogous to the construction of the ideal class group in algebraic number theory. If X is projective $C\ell(X) \simeq \mathbb{Z} \times C\ell_0(X)$, and $C\ell_0(X)$ is the group of k-rational points of an abelian variety, i.e., a group object in the category of projective varieties over k. This representability theorem fails in general if X is not projective. However we can recover representability for non projective X by refining the notion of principal divisor.

If X is not projective, let $\bar{X}$ be a smooth projective model containing X as a Zariski open set, and write $S = \bar{X} - X$; note that S is a finite set of points. Define a divisor on X to be principal "relative to S" if it is the divisor of a rational function f satisfying $f(s) = 1$ for all $s \in S$. Again, divisors principal relative to S have degree zero. We write $C\ell(\bar{X}, S)$ for the quotient of $\mathrm{Div}(X)$ by the subgroup of divisors principal relative to S, and $C\ell_0(\bar{X}, S)$ for the subgroup of divisor classes relative to S of degree 0.

Theorem . *([Se4]). With the notation above, $C\ell_0(\bar{X}, S)$ is the group of k-rational points of a connected commutative algebraic group, which is an extension of the abelian variety $\mathrm{Pic}_0(\bar{X})$ by a torus, a product of $\#(S) - 1$ copies of $GL_1 = G_m$.*

Corollary . $C\ell_0(\bar{X}, S)$ *is n-divisible if n is prime to* $\mathrm{char}(k)$.

Proof. Multiplication by n is surjective on $G_m(k) = k^*$ since k is algebraically closed while on $\mathrm{Jac}(\bar{X})$, multiplication by n is a finite morphism between irreducible varieties of the same dimension, hence surjective on k-rational points. $\square$

7.3. Generalized Jacobians and the Transfer on K-Theory

Let k, X, $\bar{X}$, S, n, be as above.

Given a point $P \in X$, the inclusion map $j_P : \mathrm{Spec}(k) \longrightarrow X$ induces a homomorphism $j_P^* : K_q(X, \mathbb{Z}/n) \longrightarrow K_q(k, \mathbb{Z}/n)$. If $D = \Sigma n_P[P] \in \mathrm{Div}(X)$, we define $j_D^* = \Sigma_P j_P^*$. Thus we obtain a pairing:

$$\mathrm{Div}(X) \otimes K_q(X, \mathbb{Z}/n) \longrightarrow K_q(k, \mathbb{Z}/n) \ .$$

Theorem . *With notation as above, if D is principal relative to S, then j_D^* is the zero map.*

Proof. Suppose $D = \operatorname{div}(f)$ for $f \in k(\bar{X})^*$ with $f(x) = 1$ for all $x \in S$. As a non zero rational function on the projective curve $\bar{X}$, f may be identified with a finite morphism $f : \bar{X} \longrightarrow \mathbb{P}^1$. We shall write $\mathbb{P}^1 - \{1\} = \mathbf{A}^1$, so $f^{-1}(\mathbf{A}^1) = U \subset X \subset \bar{X}$. Since f is finite, so is $f_U = f|_U$. Furthermore $D = \operatorname{div}(f) = f^*([0]-[\infty])$ is a divisor on U, hence $j_D^* : K_q(X, \mathbb{Z}/n) \longrightarrow K_q(k, \mathbb{Z}/n)$ factors through the restriction map $K_q(X, \mathbb{Z}/n) \longrightarrow K_q(U, \mathbb{Z}/n)$. It therefore suffices to prove the theorem when $X = U$, i.e., when $S = \{x \in \bar{X}, f(x) = 1\}$.

Then for all $t \in \mathbf{A}^1 = \mathbb{P}^1 - \{1\}$, we have a commutative diagram:

$$
\begin{array}{ccc}
f^{-1}(t) & \xrightarrow{\;j_t\;} & U \\[2mm]
f_t \;\downarrow & & \downarrow\; f \\[2mm]
\{t\} & \xrightarrow[\;i_t\;]{} & \mathbf{A}^1
\end{array}
$$

Here $f^{-1}(t)$ is a (possibly non reduced) subscheme of U. By [Q2] §7,

$$
i_t^* f_* = f_{t*} j_t^* : K_q(U, \mathbb{Z}/n) \longrightarrow K_q(k, \mathbb{Z}/n) \ .
$$

A devissage argument, c.f. [G-T], shows that $f_{t*} j_t^* = j_{f^*[t]}^*$ where $f^*[t]$ is the divisor associated to the subscheme $f^{-1}\{t\}$. Hence for $D = \operatorname{div}(f) = f^*([0] - [\infty])$, we have:

$$
j_D^* = i_0^* f_* - i_\infty^* f_* = (i_0^* - i_\infty^*) \cdot f_* \ .
$$

By the homotopy property of K-theory, $i_0^* = i_\infty^*$, hence j_D^* is the zero map. $\square$

Corollary . *(i) The pairing $\operatorname{Div}(X) \otimes K_q(X, \mathbb{Z}/n) \longrightarrow K_q(k, \mathbb{Z})$ factors through a pairing*

$$
\mu : C\ell(\bar{X}, S) \otimes K_q(X, \mathbb{Z}/n) \longrightarrow K_q(k, \mathbb{Z}/n) \ .
$$

(ii) If D is a divisor of degree zero on X, $j_D^* : K_q(X, \mathbb{Z}/n) \longrightarrow K_q(k, \mathbb{Z}/n)$ is the zero map.

Proof. i) is immediate from the definitions. To prove ii), observe by i) that this is equivalent to saying that the pairing μ vanishes on $C\ell_0(\bar{X}, S) \otimes K_q(X, \mathbb{Z}/n)$. By Theorem 7.2, $C\ell_0(\bar{X}, S)$ is n-divisible; but $K_q(X, \mathbb{Z}/n)$ is n^2-torsion, hence the pairing μ is automatically zero. $\square$

7.4. Suslin's Rigidity Theorem

Theorem (Suslin, [Su]). *Let $k \subset F$ be an extension of fields, with both k and F algebraically closed. Then the natural map $K_*(k, \mathbb{Z}/n) \longrightarrow K_*(F, \mathbb{Z}/n)$ is an isomorphism if n is prime to the characteristics of k.*

Proof. Since F is the union of its algebraically closed subfields E which have finite transcendence degree one over k,

$$BGL(F)^+ = \bigcup_E BGL(E)^+$$

and

$$K_*(F, \mathbb{Z}/n) = \varinjlim K_*(E, \mathbb{Z}/n) .$$

It therefore suffices to prove the theorem when F has finite transcendence degree over k. Using induction on $\mathrm{tr.deg}(F/k)$ we need only prove the theorem when $\mathrm{tr.deg}(F/k) = 1$. Assuming that $\mathrm{tr.deg}(F/k) = 1$, we may write

$$F = \bigcup_{R \subset F} R ,$$

as a union of regular rings which are finitely generated as k algebras. Each such R, except for k itself, is the coordinate ring of a smooth affine curve X over k.
   Since

$$K_*(F, \mathbb{Z}/n) = \varinjlim K_*(R, \mathbb{Z}/n) ,$$

to show that $K_*(k, \mathbb{Z}/n) \longrightarrow K_*(F, \mathbb{Z}/n)$ is an isomorphism, it suffices to prove, for all regular subrings $R \subset F$, finitely generated over k, that:

  i) the natural map $K_*(k, \mathbb{Z}/n) \longrightarrow K_*(R, \mathbb{Z}/n)$ is injective, and
 ii) the natural map $K_*(R, \mathbb{Z}/n) \longrightarrow K_*(F, \mathbb{Z}/n)$ factors through the map $K_*(k, \mathbb{Z}/n) \longrightarrow K_*(F, \mathbb{Z}/n)$.

Proof of i). By Hilbert's Nullstellensatz (see [La] ch. X, §2, Cor. 2) given R as above, there exists a ring homomorphism $\phi : R \longrightarrow k$, such that $\phi|_k$ is the identity. The map $K_*(k, \mathbb{Z}/n) \longrightarrow K_*(R, \mathbb{Z}/n)$ therefore has a left inverse $\phi_* : K_*(R, \mathbb{Z}/n) \longrightarrow K_*(k, \mathbb{Z}/n)$, and is injective.

Proof of ii). Writing $u : R \longrightarrow F$ and $v : k \longrightarrow R$ for the inclusions, it suffices to prove that
$$u_* = v_* \phi_* : K_*(R, \mathbb{Z}/n) \longrightarrow K_*(F, \mathbb{Z}/n) .$$

Any k-algebra homomorphism $f : R \longrightarrow F$ factors as $\tilde{f} \cdot \bar{u}$ with $\bar{u} : R \longrightarrow R \otimes_k F$ the map $\bar{u}(r) = r \otimes 1$ and $\tilde{f} : R \otimes_k F \longrightarrow F$ the map $\tilde{f}(r \otimes x) = f(r)x$. Hence $u = \tilde{u} \cdot \bar{u}$ and $v \cdot \phi = \widetilde{v \cdot \phi} \cdot \bar{u}$, so it suffices to prove that

$$\tilde{u}_* = \widetilde{v \cdot \phi}_* : K_q(R \otimes_k F) \longrightarrow K_q(F) .$$

The two F-algebra homomorphisms $\tilde{u}$ and $v \cdot \phi$ correspond to rational points x and y, respectively, of the non-singular curve $\tilde{X} = \mathrm{Spec}(R \otimes_k F)$ over F, i.e. $\tilde{u} = j_x^*$ and $\widetilde{v \cdot \phi} = j_y^*$. By theorem 7.3, $j_k^* - j_y^* = j_{[x]-[y]}^*$ is zero, since $[x] - [y]$ has degree zero, and we are done. $\square$

Corollary . *If F is an algebraically closed field, and $(n, \mathrm{char}(F)) = 1$,*

$$K_q(F, \mathbb{Z}/n) = \begin{cases} 0 & q \text{ odd} \\ \mathbb{Z}/n & q \text{ even} \end{cases} .$$

Proof. If $\mathrm{char}(F) = p > 0$, then $\bar{\mathbb{F}}_p \subset F$. By the theorem, $K_*(F, \mathbb{Z}/n) \simeq K_*(\bar{\mathbb{F}}_p, \mathbb{Z}/n)$, and Quillen's computation of $K_*(\bar{\mathbb{F}}_p, \mathbb{Z}/n)$, c.f. §7.1, finishes the proof. If $\mathrm{char}(F) = 0$, then $\bar{\mathbb{Q}} \subset F$, also $\bar{\mathbb{Q}} \subset \mathbb{C}$. Therefore $K_*(F, \mathbb{Z}/n) \simeq K_*(\mathbb{C}, \mathbb{Z}/n)$, and Suslin's computation of $K_*(\mathbb{C}, \mathbb{Z}/n)$, ([Su], c.f. Soulé's lectures) finishes the proof. $\square$

Remark(i) The corollary has also been given a direct proof by Jardine, [J], using the rigidity theorem for Henselian local rings ([G-T], Gabber (unpublished)), a result which is similar to, and inspired by, Suslin's rigidity theorem 7.4.

(ii) I have phrased the computation in terms of K-theory with coefficients. Via the long exact sequence for K-theory with coefficients, one can rephrase the result as follows:

Let F be an algebraically closed field of characteristic p. Then $K_n(F)$ is divisible for all $n > 0$, uniquely so for n even, and for n odd:

$$K_n(F)_{\mathrm{tors}} \simeq \mathbb{Q}/\mathbb{Z}[1/p] .$$

(iii) The presentation above of Suslin's result follows [G-T].

References

[Ad]   Adams, J.F.: Algebraic Topology: a students guide, London Mathematical Society Lecture Note Series, vol. 4. Cambridge Univ. Press, Cambridge, 1972

[An]   Anderson, D.W.: Relationship among K-Theories. Algebraic K-Theory I. Springer Lecture Notes in Math., vol. 341, pp. 57–72

[At]   Atiyah, M.: K-Theory. W.A. Benjamin, New York, 1967

[B-T]   Bass, H., Tate, J.: The Milnor ring of a global field. Springer Lecture Notes in Math., vol. 342, pp. 349–446

[B-F-M1] Baum, P., Fulton, W., MacPherson, R.: Riemann Roch for Singular Varieties. I.H.E.S. Pub. Math. No. 45 (1975), pp. 101–145

[B-F-M2] Baum, P., Fulton, W., MacPherson, R.: Riemann Roch and Topological K-Theory for singular varieties. Acta Math. vol. 143 (1979), pp. 155–192

[Bei]   Beilinson, A.A.: Higher Regulators and values of L-functions. J. Soviet Math. 30, 2036–2070, 1985

[Bl]   Bloch, S.: K_2 and algebraic cycles. Ann. of Math. vol. 99, 349–379, 1974

[B-S]   Borel, A., Serre, J.P.: Le Théorème de Riemann-Roch. Bull. Soc. Math.
            France, v. 86, 97–136, 1958
[B-D-F]   Brown, L.G., Douglas, R.G., Filmore, P.A.: Extensions of C^*-algebra and
            K- homology. Ann. Math. **105**, 265–324, 1977
[Bro]   Browder, W.: Algebraic K-theory with coefficients $\mathbb{Z}/p\mathbb{Z}$, Geometric appli-
            cations of homotopy theory (Proc. Conf. Evanston). Springer Lecture Notes
            vol. 657, 40–84, 1978
[Co1]   Coates, J.: K-Theory and Iwasawa's analogue of the Jacobian, Algebraic
            K-Theory II. Springer Lecture Notes in Mathematics, vol. 342, 502–520
[Co2]   Coates, J.: Research Problems: Arithmetic questions in K-theory, Algebraic
            K-theory II. Springer Lecture Notes in Math., vol. 324, 521–523
[De1]   Deligne, P.: Théorie de Hodge II. Pub. Math. I.H.E.S. vol. 40, 5–57, 1972
[De2]   Deligne, P.: Théorie de Hodge III. Pub. Path. I.H.E.S. vol. 44, 5–78, 1974
[Dou]   Douglas, R.: C^*-algebra extensions and K-homology. Annals of Math. Stud-
            ies, Princeton Univ. Press, Princeton (NJ), 1980
[Dye]   Dyer, E.: Cohomology Theories. Benjamin, Reading, Ma., 1969
[Fr]   Friedlander, E.: Etalé K-Theory I, connections with étale cohomology and
            vector bundles. Inv. Math. vol. 60, 105–134, 1980
[Gi1]   Gillet, H.: Riemann Roch Theorems for higher algebraic K-Theory. Adv. in
            Math. vol. 40, 203–289, 1981
[Gi2]   Gillet, H.: Comparison of K-theory spectral sequences with applications. In:
            Algebraic K-Theory. Evanston, 1980. Springer Lecture Notes in Mathemat-
            ics vol. 854, 141–167, 1981
[Gi3]   Gillet, H.: K-Theory and intersection theory, revisited K-theory, vol. 1, 405–
            415, 1987
[G-G]   Gillet, H., Grayson, D.: The loop space of the Q construction. Illinois Journal
            of Math. vol. 31, 574–597, 1987
[G-T]   Gillet, H., Thomason, R.: The K-theory of strict Hensel local rings and a
            theorem of Suslin. Journal of Pure and Applied Algebra, vol. 34, 241–254,
            1984
[Go]   Godement, R.: Topologie Algébrique et théorie des Faisceaux. Hermann,
            Paris, 1958
[Gr1]   Grayson, D.: Higher Algebraic K-Theory II (after Daniel Quillen), Algebraic
            K-theory (proc. Conf. Northwestern Univ. 1976). Springer Lecture Notes in
            Math. vol. 551, 217–240, 1976
[Gr2]   Grayson, D.: Products in K-Theory and intersecting algebraic cycles. Inv.
            Math. vol. 47, 71–83, 1978
[G-A]   Griffiths, P., Adams, J.: Topics in algebraic and analytic geometry. Mathe-
            matical Notes, vol. 13, Princeton Univ. Press. Princeton, N.J., 1974
[G-H]   Griffiths, P., Harris, J.: Principles of algebraic geometry. John Wiley and
            Sons, New York, 1978
[Ha]   Hartshorne, R.: Algebraic Geometry. Graduate Texts in Mathematics, vol.
            52. Springer Verlag, New York, 1977
[Har]   Hartshorne, R.: Equivalence relations in algebraic cycles. In: Proc. Symp.
            Pure Math., vol. 29, 1974
[Hi]   Hironaka, H.: Triangulations of algebraic sets. Proceedings of Symposia in
            Pure Math., vol. 29, 165–185, 1975
[Hi]   Hirzebruch, F.: Topological methods in algebraic geometry, third edition.
            Grundlehren der mathematischen Band 131. Springer Verlag, New York,
            1966

[J]   Jardine, J.F.: Simplicial objects in a Grothendieck topos. Contemporary Math., vol. 55 (pt 1), 193–240, 1986

[Ka1]   Karoubi, M.: K-Theory, an introduction. Grundlehren der mathematischen Wissenschaften, band 226. Springer Verlag, Berlin, New York, 1978

[Ka2]   Karoubi, M.: Homologie Cyclique et K-Théorie, Asterisque vol. 149, 1987

[La]   Lang, S.: Algebra. Addison-Wesley, 1965

[L]   Lichtenbaum, S.: Values of zera-functions, étale cohomology, and algebraic K-theory. Algebraic K-Theory II. Springer Lecture Notes in mathematics, vol. 342, pp. 489–501

[Mi]   Milnor, J.: Introduction to Algebraic K-Theory. Annals of math. Studies, No. 72, Princeton Univ. Press, Princeton, 1971

[Mi-S]   Milnor, J., Stasheff, J.: Characteristic classes. Annals of Mathematics Studies, No. 76. Princeton University Press, Princeton, N.J., 1974

[Mu]   Mumford, D.: Rational equivalence of O-cycles on surfaces. J. Math. Kyoto Univ. 9, 105–204, 1968

[N]   Neisendorfer, J.: Primary homotopy theory. Memoirs AMS, vol. 232, 1980

[Q1]   Quillen, D.: On the cohomology and K-theory of the general linear group over a finite field. Ann. of Math., vol. 96, 552–586, 1972

[Q2]   Quillen, D.: Higher Algebraic K-Theory I. Springer Lecture Notes in Math., vol. 341, pp. 85–147

[Q3]   Quillen, D.: Higher Algebraic K-Theory. Proceedings of the International Congress of Mathematicians, Vancouver, vol. 1, 171–176, 1974

[R-S-S]   Rapoport, M., Schappacher, N., Schneider, P. (eds.): Beilinson's conjectures on special values of L-functions. Perspectives in Math., vol. 4. Academic Press, San Diego, 1988

[SGA6]   Seminaire de Geometrie Algébrique et Théorème de Riemann-Roch (Dirigé par P. Berthelot, A. Grothendieck et L. Illusie). Springer Lecture Notes in Math. vol. 225

[Se1]   Serre, J.-P.: Faisceaux algébriques cohérents. Ann. of Math., vol. 61, 197–278, 1955

[Se2]   Serre, J.-P.: Géométrie algébrique et géométrie analytique. Ann. Inst. Fourier, vol. 6, 1–42, 1956

[Se3]   Serre, J.-P.: Modules projectifs et espaces fibrés à fibre vectorielle. Sem. Dubreil, No. 23, 1957–1958

[Se4]   Serre, J.P.: Groupes algébriques et corps de classes. Hermann, Paris, 1959

[Sou]   Soulé, C.: Opérations en K-théorie algébrique. Canadian J. Math. vol. 37, 488–550, 1985

[Sp]   Spanier, E.: Algebraic Topology. McGraw Hill, New York, 1966

[Su]   Suslin, A.A.: On the K-Theory of algebraically closed fields. Inv. Math. 73, 241–245, 1983

[Sw]   Swan, R.: Vector Bundles and Projective Modules. Trans. Am. Math. Soc. vol. 105, 264–277, 1962

[Th]   Thomason, R.: Algebraic K-Theory and étale cohomology. Ann. Scient. Ec. Norm. Sup. 4e série, t. 13, 437–552, 1980

[T2]   Thomason, R.: The Lefschetz-Riemann-Roch Theorem and the coherent trace formula. Inv. Math. 85, 516–543, 1986

[W]   Waldhausen, F.: Algebraic K-theory of generalized free products. Ann. of Math. vol. 108, 135–136, 1978

Algebraic K-Theory of the Integers

Christophe Soulé

Institut des Hautes Études Scientifiques, 35, Route de Chartres, F-91440 Bures-sur-Yvette

Since the very definition of higher algebraic K-theory by Quillen, it has been a challenge to compute the higher K-theory of the ring $\mathbb{Z}$ of integers. It is still unknown whether the conjecture of Lichtenbaum is true, which says that the quotient of the order of the torsion of two consecutive K-groups of $\mathbb{Z}$ is equal to a Bernoulli number. More generally, algebraic K-groups of any arithmetic variety are expected to be related to the values of its zeta function at integral points.

In Chapter I we describe known results about the K-theory of $\mathbb{Z}$ and explain the classical fact that Bernoulli numbers are the values of the Riemann zeta function at odd negative integers. In Chapter II we sketch a proof of the finite generation of each K-group of $\mathbb{Z}$, using the reduction of quadratic forms. In Chapter III we compute the K-theory with finite coefficients of the complex numbers; the reader is referred to Gillet's lectures in this volume for the extension of the result to arbitrary algebraically closed fields. In Chapter IV we study the Chern classes of the standard representation of $GL_N(\mathbb{C})$ and $GL_N(\mathbb{Z})$. In Chapter V we describe briefly how Bernoulli numbers are related to the K-theory of $\mathbb{Z}$ via its link to the étale cohomology of $\mathrm{Spec}(\mathbb{Z}[1/p])$.

The reader might be disappointed by the fact that many proofs will be skipped. We still hope that he will get some feeling of the methods that have been used so far to compute the K-theory of the integers, and their link to several aspects of number theory.

I would like to thank E. Lluis-Puebla and the Universidad Nacional Autónoma de México for organizing this series of talks and for their help in the preparation of these notes.

I. Introduction: K-Theory of Integers and Values of Zeta Functions

1. K-Theory of $\mathbb{Z}$

1.1. Given a ring A with unit, one can define a series of abelian groups $K_m(A)$, $m \in \mathbb{N}$, called the higher algebraic K-theory groups of A. When $m = 0$, $K_0(A)$ is the Grothendieck groups of the category $P(A)$ of finitely generated projective A-modules. This means that $K_0(A)$ is generated by elements $[P]$, where P runs over all objects in $P(A)$, with relations $[P] = [P'] + [P'']$ whenever there exists an exact sequence

$$0 \longrightarrow P' \longrightarrow P \longrightarrow P' \longrightarrow 0$$

in $P(A)$. For instance, when $A = \mathbb{Z}$, the map which sends a module P to its rank induces an isomorphism $K_0(\mathbb{Z}) \cong \mathbb{Z}$ (all projective $\mathbb{Z}$-modules are free).

When $m > 0$, a definition of $K_m(A)$ is ([35], [27])

$$K_m(A) = \pi_m BGL(A)^+ \ ,$$

where $BGL(A)^+$ is a topological space obtained by adding cells in dimension 2 and 3 to $BGL(A)$ in order to make it an H-group without changing its homology with locally constant coefficients. Here $BGL(A)$ denotes the classifying space of the group of infinite matrices

$$GL(A) = \varinjlim_{N \geq 1} GL_N(A) \ ,$$

where $GL_N(A)$ imbeds in $GL_{N+1}(A)$ by sending g to

$$\begin{bmatrix} & & & 0 \\ & g & & \vdots \\ & & & 0 \\ 0 & \ldots & 0 & 1 \end{bmatrix} .$$

In particular $H_m(BGL(A)^+; \mathbb{Z}) = H_m(GL(A); \mathbb{Z})$ for any integer $m \geq 0$ (for more details see Chapter II and Loday's lectures).

1.2. In these lectures we shall be interested in studying the algebraic K-theory of the ring $\mathbb{Z}$, and more generally of the ring of algebraic integers in a number field F. Recall that a number field F is a finite extension of $\mathbb{Q}$, i.e. any element x in F satisfies an algebraic equation

$$x^n + a_1 x^{n-1} + \ldots + a_n = 0 \tag{$*$}$$

with coefficients $a_1, \ldots, a_n$ in $\mathbb{Q}$. The ring of integers in F consists of the elements x in F which satisfies an equation $(*)$ where all coefficients $a_1, \ldots, a_n$ lie in $\mathbb{Z}$.

Theorem 1.1 [36]. *When A is a ring of algebraic integers and $m \geq 0$, the group $K_m(A)$ is finitely generated.*

Theorem 1.2 [8]. *Let F be a number field, r_1 (resp. r_2) its number of real (resp. non conjugate complex) imbeddings, and A its ring of integers. Then $K_m(A) \otimes \mathbb{Q}$ is trivial for $m > 0$ even, it is isomorphic to $\mathbb{Q}^{r_1+r_2-1}$ when $m = 1$, to $\mathbb{Q}^{r_2}$ when $m = 3 + 4m'$, and to $\mathbb{Q}^{r_1+r_2}$ when $m = 5 + 4m'$.*

For instance $K_m(\mathbb{Z})$ is the sum of $\mathbb{Z}$ with a finite group when $m = 0, 5, 9, 13, 17$, etc. ...; it is finite otherwise. Therefore we are faced with the following:

Problem 1.2. What is the order of the torsion subgroup of $K_m(\mathbb{Z})$?.

2. Bernoulli Numbers

2.1. Let us consider the function

$$F(t) = \frac{te^t}{e^t - 1} \ .$$

We develop it as a power series in the variable t:

$$F(t) = \sum_{n=0}^{\infty} b_n \frac{t^n}{n!} \ .$$

The numbers b_n are rational numbers. These are called the *Bernoulli numbers*.
  One computes

$$b_0 = 1, b_1 = 1/2, b_2 = 1/6, b_3 = 0, b_4 = -1/30, \ldots,$$
$$b_{12} = -691/2730, \ldots \ .$$

Since $F(t) - \frac{t}{2}$ is an even function, $b_n = 0$ when n is odd and bigger than one:

$$F(-t) = \frac{-te^{-t}}{e^{-t} - 1} = \frac{-t}{1 - e^t} = -t - \frac{te^t}{1 - e^t} = -t + F(t) \ .$$

2.2. It is expected that Bernoulli numbers will come in when answering the Problem 1.2.

Conjecture 2.2 [26]. *Assume $n > 0$ is even. Then*

$$\frac{\# K_{2n-2}(\mathbb{Z})}{\# K_{2n-1}(\mathbb{Z})} = \left| \frac{b_n}{n} \right| 2^r$$

for some integer $r \in \mathbb{Z}$.

Here $\# X$ denotes the cardinality of a finite set X.

2.3. The groups $K_m(\mathbb{Z})$ for small values of m can be computed by studying the homology of the arithmetic groups $GL_N(\mathbb{Z})$ (see also Chapter II, 6). This gives

$$K_1(\mathbb{Z}) = \mathbb{Z}/2 \quad [32],$$
$$K_2(\mathbb{Z}) = \mathbb{Z}/2 \quad [32],$$
$$K_3(\mathbb{Z}) = \mathbb{Z}/48 \quad [23],$$
$$K_4(\mathbb{Z}) = 2 - \text{torsion} \oplus (\mathbb{Z}/3 \text{ or } 0), \quad [24], [40],$$
$$K_5(\mathbb{Z}) = 2 - \text{torsion} \oplus 3 - \text{torsion} \oplus \mathbb{Z}, \quad [24] .$$

So, when $n = 2$, in Conjecture 2.2, we get

$$\frac{\# K_2(\mathbb{Z})}{\# K_3(\mathbb{Z})} = \frac{2}{48} = \frac{1}{24}$$

when $\frac{b_2}{2} = \frac{1}{6 \cdot 2} = \frac{1}{12}$.

Further evidence for Conjecture 2.2 is given by:

Theorem 2.3. *[37], [41], [16]. Let $n > 0$ be an even integer. The denominator of b_n/n divides $\# K_{2n-1}(\mathbb{Z})$, and the numerator of $\frac{b_n}{n}$ divides $\# K_{2n-2}(\mathbb{Z})$.*

For example, let $n = 12$. It is known that

$$b_{12}/12 = -691/32760$$

(12 is the smallest value of n for which the numerator of b_n/n is not equal to one). Theorem 2.3 implies that $K_{22}(\mathbb{Z})$ has an element of order 691.

3. Values of the Riemann Zeta Function

3.1. When a complex number s has real part bigger than one, the series

$$\zeta(s) = \sum_{m \geq 1} m^{-s}$$

converges. This is the *Riemann zeta function*.

On the other hand, when $\text{Re}(s) > 0$, the integral

$$\Gamma(s) = \int_0^\infty e^{-t} t^{s-1} dt$$

converges. This is the *gamma function*.

Let us state some (classical) properties of these functions:

Theorem 3.1. *i) $\Gamma(s)$ has a meromorphic continuation to the whole complex plane.*
ii) $\Gamma(s+1) = s\Gamma(s)$. $\Gamma(n) = (n-1)!$ when n is a positive integer.
iii) $\Gamma(s)$ has a pole of order 1 at $s = 0, -1, -2, -3, \ldots$.
iv) $\sin(\pi s)\Gamma(s)\Gamma(1-s) = \pi$.

Theorem 3.2. *i) $\zeta(s)$ has a meromorphic continuation to the whole complex plane.*

ii) When $n \geq 1$, $\zeta(1 - n) = -\frac{b_n}{n}$.

iii) Let $\xi(s) = \pi^{-\frac{s}{2}} \Gamma\left(\frac{s}{2}\right) \zeta(s)$. *Then*

$$\xi(s) = \xi(1 - s) .$$

iv) The function $\zeta(s)$ *has a simple pole at* $s = 1$, *it is holomorphic when* $s \neq 1$.
v) When $n > 1$ *is odd,* $\zeta(s)$ *has a simple zero at* $s = 1 - n$.

3.2. Let us give a partial sketch of the proof of Theorem 3.1. If $\mathrm{Re}(s) > 0$ we get, by integrating by parts,

$$\Gamma(s + 1) = \int_0^\infty e^{-t} t^s \, dt = - \int_0^\infty \left(e^{-t}\right)' t^s \, dt$$

$$= \int_0^\infty e^{-t} (t^s)' \, dt - \left[e^{-t} t^s\right]_0^\infty$$

$$= s \int_0^\infty e^{-t} t^{s-1} \, dt = s\Gamma(s) .$$

Furthermore $\Gamma(1) = \int_0^\infty e^{-t} dt = 1$, therefore, if $n > 0$ is an integer, we get, by induction on n,

$$\Gamma(n) = (n - 1)\Gamma(n - 1) = (n - 1)(n - 2)! = (n - 1)! .$$

To prove i), let $k \in \mathbb{N}$ and $s \in \mathbb{C}$ with $-k < \mathrm{Re}(s) \leq -k + 1$. By induction on k we define $\Gamma(s) = \frac{\Gamma(s+1)}{s}$.

This also proves ii) when $\mathrm{Re}(s) \leq 0$ and iii) by induction on $k = -s$. We shall not prove iv).

3.3. To prove Theorem 3.2, let us introduce, for any complex variable z, the functions

$$F(z) = \frac{z e^z}{e^z - 1}$$

and

$$G(z) = \frac{e^{-z}}{1 - e^{-z}} .$$

We have $G(t) = \sum_{n \geq 1} e^{-nt}$ for $t > 0$, therefore, when $\mathrm{Re}(s) > 1$,

$$\int_0^\infty G(t) t^{s-1} \, dt = \int_0^\infty \sum_{n=1}^\infty e^{-nt} t^{s-1} \, dt$$

$$= \sum_{n=1}^\infty \int_0^\infty e^{-nt} t^{s-1} \, dt$$

$$= \sum_{n=1}^\infty n^{-s} \int_0^\infty e^{-u} u^{s-1} \, du = \zeta(s)\Gamma(s) .$$

If $z \in \mathbb{C}^*$, let $\log(z)$ be the principal value of the logarithm. Namely, write $z = re^{i\theta}$, $r > 0$, $-\pi < \theta \le \pi$. Then $\log(z) = \log(r) + i\theta$. We also define $z^{s-1} = e^{(s-1)\log(z)}$. Consider, for any small real number $\varepsilon > 0$, the oriented contour of integration $C_\varepsilon = (-\infty, -\varepsilon) \cup K_\varepsilon \cup (-\varepsilon, -\infty)$, where K_ε is the circle $K_\varepsilon = \{z / |z| = \varepsilon\}$:

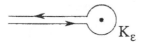

On $(-\infty, -\varepsilon)$ (resp. $(-\varepsilon, -\infty)$) we have $\log(z) = \log|z| - \pi i$ (resp. $\log(z) = \log|z| + \pi i$) .

Let

$$H(s) = \int_{C_\varepsilon} F(z) z^{s-1} \frac{dz}{z} \ .$$

This function is well defined for any complex number s (when it goes to $-\infty$, the exponential e^{-t} dominates $e^{s \log t}$). Furthermore, by changing z to $-z$ and since $F(-z) = zG(z)$, we get

$$H(s) = \int_{-C_\varepsilon} F(-z)(-z)^{s-1} \frac{dz}{z} = -\int_{-C_\varepsilon} G(z)(-z)^s \frac{dz}{z}$$

$$= e^{-\pi i s} \int_\varepsilon^\infty G(t) t^{s-1} dt + \int_{-K_\varepsilon} G(z)(-z)^{s-1} dz$$

$$- e^{\pi i s} \int_\varepsilon^\infty G(t) t^{s-1} dt \ .$$

Assume that $\mathrm{Re}(s) > 1$. Then the integral

$$\int_{-K_\varepsilon} G(z)(-z)^{s-1} dz$$

tends to zero with ε. Therefore we get

$$H(s) = -\left(e^{\pi i s} - e^{-\pi i s}\right) \int_0^\infty G(t) t^{s-1} dt$$

$$= -2i \sin(\pi s) \Gamma(s) \zeta(s) = -\frac{2\pi i}{\Gamma(1-s)} \zeta(s)$$

(by Theorem 3.1. iv). Since $H(s)$ is holomorphic and $\Gamma(1 - s)$ is meromorphic (Theorem 3.1. i)), this formula proves Theorem 3.2. i) . Furthermore, since $\zeta(s)$ has no pole when $\mathrm{Re}(s) > 1$, we deduce, from Theorem 4 iii), that $\zeta(s)$ can have a pole only when $s = 1$. The residue of $\zeta(s)$ at $s = 1$ is not zero since

$$H(1) = \int_{K_\varepsilon} F(z) \frac{dz}{z} = 2\pi i \ .$$

This proves Theorem 3.2. iv).

Let us prove Theorem 3.2. ii). We have

$$\zeta(1-n) = -\frac{\Gamma(n)}{2\pi i} H(1-n)$$

where

$$H(1-n) = \int_{C_\epsilon} F(z) z^{-n} \frac{dz}{z} = \int_{K_\epsilon} F(z) z^{-n} \frac{dz}{z}$$

(since z^{-n} is well defined on $\mathbb{C}^*$). Thus

$$H(1-n) = \int_{K_\epsilon} \left(\sum_{m \geq 1} b_m \frac{z^m}{m!} \right) z^{-n} \frac{dz}{z} = 2\pi i \frac{b n}{n!} \ .$$

Thus, since $\Gamma(n) = (n-1)!$ (Theorem 3.1, ii)) we get

$$\zeta(1-n) = -\frac{b_n}{n} \ .$$

We shall not prove Theorem 3.2, iii). Let us just indicate that it can be obtained by integrating $F(z) z^{s-1} \frac{dz}{z}$ along the following type of paths:

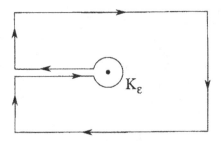

More details can be found for example in [22], Appendix (that we followed for the proof above).

Finally, Theorem 3.2 v), follows from Theorem 3.2 iii) and Theorem 3.1 iii), since $\zeta(n) \neq 0$ when $n > 0$.

3.4. By Theorem 3.2, ii), Conjecture 2.2 is equivalent to

$$\frac{\# K_{2n-2}(\mathbb{Z})}{\# K_{2n-1}(\mathbb{Z})} = |\zeta(1-n)| 2^r \ .$$

This is the prototype of an impressive set of conjectures relating the K-theory of an arithmetical variety to the values of its zeta function at integral points [26], [6], [5], [44]. For instance, given a number field F, one defines its zeta function by

$$\zeta_F(s) = \sum_a (Na)^{-s}, \quad \text{if} \quad \mathrm{Re}(s) > 1 \ ,$$

where a runs over all non zero ideals in the ring A of integers in F and $Na = \#(A/a)$. This function can be analytically continued to the whole complex plane. When F is totally real ($r_2 = 0$) and $n \geq 2$ is even it is known that $\zeta_F(1-n)$ is a rational number. On the other hand $K_{2n-2}(A)$ and $K_{2n-1}(A)$ are finite groups (see 1.2.). Lichtenbaum conjectured [26] that

$$\frac{\# K_{2n-2}(A)}{\# K_{2n-1}(A)} = \zeta_F(1-n)2^r$$

for some integer $r \in \mathbb{Z}$.

II. Finite Generation of the K-Theory of Integers

1. The K-theory of $\mathbb{Z}$ (bis)

1.1. Let G be a discrete group and EG be a space (by this we mean a CW complex) on which G acts freely (by cellular homeomorphisms) and which is contractible. The quotient space $BG = EG/G$ is called the *classifying space* of G. It is well defined up to homotopy. One way to make it functorial is to define EG as the geometric realization of the simplicial set equal to G^{n+1} in degree n, with faces (resp. degeneracies) given by forgetting (resp. repeating) one element. The action of G is given by translation of each coordinate.

By definition, the homology of BG is the homology of the group G, i.e. $H_*(G, \mathbb{Z}) = H_*(BG, \mathbb{Z})$. As far as homotopy is concerned, BG is rather trivial: it is connected, its fundamental group $\pi_1(BG)$ is equal to G, and $\pi_m(BG) = 0$ when $n > 1$ (in other words BG is a $K(G, 1)$).

1.2. An *H-group* X is a space X equipped with maps

$$\mu : X \times X \longrightarrow X, \quad i : X \longrightarrow X$$

and a point $e \in X$ which satisfy the axioms of a group up to homotopy. That is, the maps $X^3 \longrightarrow X$ defined by sending (x, y, z) to $\mu(x, \mu(y, z))$ or $\mu(\mu(x, y), z)$ are homotopic, the maps $X \longrightarrow X$ sending x to $\mu(e, x)$ and $\mu(x, e)$ are homotopic to the identity, and the map $X \longrightarrow X$ sending x to $\mu(x, i(x))$ (or $\mu(i(x), x)$) is homotopic to the constant map sending x to e.

The fundamental group of an H-group is automatically abelian. The Hurewicz map $h : \pi_1(X) \longrightarrow H_1(X, \mathbb{Z})$ is an isomorphism.

1.3. Let A be an associative ring with unit, $GL_N(A)$ the group of invertible N by N matrices with coefficients in A,

$$GL(A) = \varinjlim_{N \geq 1} GL_N(A)$$

and $BGL(A)$ the classifying space of $GL(A)$. Quillen defined a space $BGL(A)^+$ (by adding cells in dimensions two and three) such that:

i) $BGL(A)^+$ is a connected H-group;
ii) The map φ induces an isomorphism in homology with (locally) constant coefficients.

In other words, if M is a locally constant sheaf of abelian groups over $BGL(A)^+$ the natural map $\varphi_* : H_m(BGL(A), \varphi^*(M)) \longrightarrow H_m(BGL(A)^+, M)$ is an isomorphism (see [34] and [27], and Loday's Lecture No. 5).

Definition . $K_m(A) = \pi_m(BGL(A)^+)$ when $m > 0$.

For example,

$$K_1(A) = \pi_1(BGL(A)^+) \xrightarrow{\hbar} H_1(BGL(A)^+, \mathbb{Z}) = H_1(GL(A), \mathbb{Z})$$

is the quotient $GL(A)/[GL(A), GL(A)]$ of $GL(A)$ by its commutator subgroup.

2. Finite generation

2.1. We shall describe a proof of

Theorem 2.1. *[36]. For any $m \geq 1$ the group $K_m(\mathbb{Z})$ is finitely generated.*

Given a connected CW complex X, the Hurewicz theorem modulo the Serre category of finitely generated groups states that the following statements are equivalent:

i) For all $m \geq 1$, $\pi_m(X)$ is finitely generated;
ii) For all $m \geq 2$, $H_m(X, \mathbb{Z})$ is finitely generated, and $\pi_1(X)$ is finitely generated.

Therefore Theorem 1 is equivalent to the finite generation of the homology groups $H_m(GL(\mathbb{Z}), \mathbb{Z})$, $m \geq 1$.

2.2. The first step is a stability argument which, for m fixed, makes it possible to replace $GL(\mathbb{Z})$ by a given $GL_N(\mathbb{Z})$.

For this we just quote a general result of stability.

Theorem 2.2 [46]. *Let A be any noetherian commutative ring with unit. Assume that the Krull dimension $\dim(A)$ is finite. Then the natural map $H_m(GL_N(A), \mathbb{Z}) \longrightarrow H_m(GL(A), \mathbb{Z})$ is an isomorphism when $N \geq \mathrm{Max}(2m + 1, m + \dim(A) - 1)$.*

Recall that the Krull dimension of A is the maximum of the set of integers n such that there exists a sequence

$$\wp_0 \subsetneq \wp_1 \subsetneq \wp_2 \subsetneq \ldots \subsetneq \wp_n$$

of prime ideals in A (Corollary 8.3. in loc. cit. is actually stronger and involves the stable range of A; this number is less or equal to $\dim(A) - 1$).

Since $\mathbb{Z}$ has dimension one we get

$$H_m(GL_N(\mathbb{Z}), \mathbb{Z}) \xrightarrow{\sim} H_m(GL(\mathbb{Z}), \mathbb{Z})$$

when $N \geq 2m + 1$. Therefore Theorem 2.1 will follow from

Theorem 2.3. *For every $N \geq 1$, and $m \geq 0$, the group $H_m(GL_N(\mathbb{Z}), \mathbb{Z})$ is finitely generated.*

To prove this we shall use the reduction theory of quadratic forms on $\mathbb{R}^N$.

3. Reduction Theory

3.1. We fix an integer $N \geq 1$. The arithmetic groups first appeared in the work of mathematicians like Gauss and Minkowski, when they studied the *reduction theory of quadratic forms*.

Consider a quadratic form in N-variables

$$\sum_{1 \leq i,j \leq N} a_{ij} x_i x_j$$

where a_{ij} are integers and $a_{ij} = a_{ji}$.

Problem 3.1. Given $y \in \mathbb{Z}$ can one find $x_i \in \mathbb{Z}$, $1 \leq i \leq N$, such that

$$\sum_{1 \leq i,j \leq N} a_{ij} x_i x_j = y \ ?$$

If this problem has an affirmative answer we shall say that y *is h-representable*, where h is the square matrix $h = (a_{ij})$.

3.2. The group $GL_N(\mathbb{Z})$ is a group of symmetry for the Problem 3.1. Indeed let v be the column vector (x_i) and $g \in GL_N(\mathbb{Z})$. Consider the vector $v' = g^{-1}(v) = (x_i')$. Let

$$h[v] = \sum_{i,j} a_{ij} x_i x_j = (v^t) \, h v \ ,$$

where v^t is the transpose of v. We get

$$h[v] = h[g(v')] = (gv')^t g(hv') = (v')^t (g^t) h g v'$$
$$= (g \cdot h)[v']$$

where $g \cdot h$ is, by definition, the matrix $g \cdot h = (g^t) h g$. We have thus proved the following

Lemma 3.2. *The number y is h-representable if and only if it is $g \cdot h$-representable.*

In order to study Problem 3.1 it is therefore natural first to describe the set of orbits of $GL_N(\mathbb{Z})$ in the space of (positive definite) real valued quadratic forms. Such a description is the purpose of reduction theory (see [7]). This is related to Theorem 1 by giving us a concrete realization of the classifying space of $GL_N(\mathbb{Z})$, or, at least, of a congruence subgroup in $GL_N(\mathbb{Z})$.

4. Congruence Subgroups

4.1. Let $\Gamma \subset GL_N(\mathbb{Z})$ be the congruence subgroup modulo 3, i.e. the group Γ made of matrices $g \in GL_N(\mathbb{Z})$ which can be written as $g = 1 + 3m$, where $m \in M_N(\mathbb{Z})$ is any N by N square matrix.

Lemma 4.1. *The group Γ has no nontrivial element of finite order.*

Proof. For any $n \geq 1$ let $U_n \subset \Gamma$ be the subgroup made of elements g of the form $g = 1 + 3^n m$, $m \in M_N(\mathbb{Z})$. Assume $g \in U_n - U_{n+1}$. Then 3 dies not divide m and $g^2 = 1 + 3^n 2m + 3^{2n} m^2$ is in $U_n - U_{n+1}$, when $g^3 = 1 + 3^{n+1} m + 3^{2n+1} m^2 + 3^{3n} m^3$ is in $U_{n+1} - U_{n+2}$. More generally, let $k \in \mathbb{N}$ be written as $k = 3^\alpha k'$, where 3 does not divide k'. Then $g^k \in U_{n+\alpha} - U_{n+\alpha+1}$. Therefore $g^k \neq 1$. $\qquad\square$

4.2. Let us consider the space $\mathcal{H}$ of all positive definite real quadratic forms on $\mathbb{R}^N$. An element $h \in \mathcal{H}$ is a matrix $h = (a_{ij})$, $1 \leq i, j \leq N$, $a_{ij} \in \mathbb{R}$, $a_{ij} = a_{ji}$, and, for all real vectors v, $h[v] \geq 0$, with equality only when $v = 0$. The group R_+^* acts upon $\mathcal{H}$ by multiplication of the matrix h by $\lambda \in \mathbb{R}_+^*$. We call $X = \mathcal{H}/\mathbb{R}_+^*$ the quotient space. The action of $GL_N(\mathbb{Z})$ on $\mathcal{H}$ is given by $g \cdot h = (g^t)hg$ as in 3.2. This action commutes with R_+^*, therefore $GL_N(\mathbb{Z})$ acts upon X.

Lemma 4.2. *i) X is contractible.*
ii) Γ acts freely upon X.

Proof. i) Given $h \in \mathcal{H}$ and $t \in [0,1]$ let

$$\varphi_t(h) = \big(th + (1-t)Id\big) \ ,$$

where Id is the N by N identity matrix. We have $\varphi_1(h) = h$ and $\varphi_0(h) = Id$. Furthermore $\varphi_t(h)$ is in $\mathcal{H}$: for any vector $v \in \mathbb{R}^N$ we have

$$\varphi_t(h)[v] = th[v] + (1-t)|v|^2 \geq 0 \ ,$$

with equality only when $v = 0$. Here

$$|v|^2 = \sum_{i=1}^{i=N} x_i^2$$

denotes the standard norm on $\mathbb{R}^N$. This proves that $\mathcal{H}$ is contractible.

This contraction is not $\mathbb{R}_+^*$-equivariant but we can, for instance, identify X which the subset $\mathcal{H}_1$ in $\mathcal{H}$ made of matrices h with determinant 1. The family of maps $\psi_t(h) = \varphi_t(h) \det\big(\varphi_t(h)\big)^{-1}$, $t \in [0,1]$, defines a contraction of $\mathcal{H}_1$ to $Id \in \mathcal{H}_1$.

ii) Let $g \in GL_N(\mathbb{Z})$ and $h \in \mathcal{H}$ be such that $g \cdot h = h$. We shall see that g has finite order. For any v in $\mathbb{R}^N$ we have $g \cdot h[v] = h[v]$, i.e. $h[g(v)] = h[v]$ (cf. 3.2.). Let (e_i) be the standard basis of $\mathbb{R}^N$ and $K = \sup_{1 \leq i \leq N} h[e_i]$. Let $C \subset \mathbb{R}^N$ be the compact set made of vectors v such that $h[v] \leq K$. Since

$h[g(e_i)] = h[e_i] \leq K$, all the columns of g belong to C. Since $\mathbb{Z}^N$ is discrete in $\mathbb{R}^N$, the set $\mathbb{Z}^N \cap C$ is finite, therefore there are only finitely many $g \in GL_N(\mathbb{Z})$ such that $g \cdot h = h$. These form a group, therefore they have finite order.

Now Γ has no torsion (4.1.), so it acts freely on $\mathcal{H}$. Finally, let $\lambda \in \mathbb{R}_+^*$, $h \in \mathcal{H}$ and $g \in \Gamma$ be such that $g \cdot h = \lambda h$. Then $\lambda^N \det(h) = \det(g.h) = \det(h)$, so $\lambda = 1$. This proves ii).

4.3. The quotient $G = GL_N(\mathbb{Z})/\Gamma = GL_N(\mathbb{F}_3)$ is a finite group. The homology of $GL_N(\mathbb{Z})$ can be computed from the action of G on the homology of Γ via the Hochschild-Serre spectral sequence

$$E_{pq}^2 = H_p(G, H_q(\Gamma, \mathbb{Z})) \Rightarrow H_{p+q}(GL_N(\mathbb{Z}), \mathbb{Z}) \ .$$

Therefore, to prove that each group $H_m(GL_N(\mathbb{Z}), \mathbb{Z})$, $m > 0$, is finitely generated, it is enough to prove that $H_m(\Gamma, \mathbb{Z})$ is finitely generated for all $m > 0$.

5. A Good Subspace

5.1. In view of 4.2., if we admit that X has a cell decomposition invariant by Γ (as we shall!), we see that X/Γ is homotopic to $B\Gamma$. To prove that Γ has finitely generated homology in each degree we need to show this for X/Γ. If X/Γ was compact we would be done. However this is false for all $N > 1$ (X/Γ has only finite volume, see [7]).

One way to overcome this difficulty is to extend the action of Γ to a space $\bar{X}$ (which is still contractible) whose quotient by Γ is a compactification of X/Γ [10]. Alternatively, one can define $Y \subset X$ such that: Y is a retract of X, Y is stable by Γ, and Y/Γ is compact. Such a "good subspace" is described below.

5.2. Let $h \in \mathcal{H}$ be a quadratic form. Since for any $K \in \mathbb{R}_+^*$ the set of $v \in \mathbb{Z}^N$ such that $h[v] \leq K$ is finite (cf. 4.2.), h has a finite set of nonzero minimal vectors $\mathrm{Min}(h) \subset \mathbb{Z}^N$. Let H be the set of $h \in \mathcal{H}$ such that $\mathrm{Min}(h)$ generates the real vector space $\mathbb{R}^N$.

This set H is stable by the action of $\mathbb{R}_+^*$ $\big(\mathrm{Min}(\lambda h) = \mathrm{Min}(h)\big)$ and by the action of $GL_N(\mathbb{Z})\big(\mathrm{Min}(g \cdot h) = g\big(\mathrm{Min}(h)\big)\big)$. Let $Y = H/\mathbb{R}_+^*$. The following was proved by J. Lannes and myself [40], and extended by A. Ash [3] to a more general situation:

Proposition 5.2. *i) There exists a $GL_N(\mathbb{Z})$-equivariant retraction from X to Y.*
ii) $Y/GL_N(\mathbb{Z})$ (hence Y/Γ) is compact.

Using this we conclude that $B\Gamma = Y/\Gamma$ has finitely generated homology in each degree, and, by 4.3., this concludes the proof of Theorem 1.

5.3. We shall not prove the Proposition 5.2. but just give the example $N = 2$. Then X is the Poincaré upper half plane $X = \{z \in \mathbb{C}/\mathrm{Im}(z) > 0\}$. The group

$GL_2(\mathbb{Z})$ (or rather $SL_2(\mathbb{Z})$) acts upon X by homographic transformations with fundamental domain the set F of $z = x + iy$ such that $-\frac{1}{2} \leq x \leq \frac{1}{2}$ and $x^2 + y^2 \geq 1$. The subset Y is the tree introduced by Serre in [38]. Namely Y is the union of the translates by $SL_2(\mathbb{Z})$ of $C = \{z \in F/x^2 + y^2 = 1\}$.

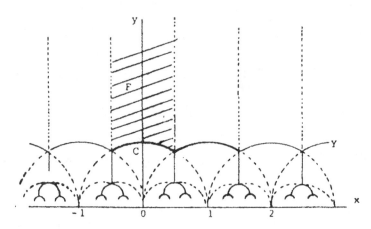

6. Conclusion

The reduction theory of quadratic forms, i.e. the definition of a fundamental domain for the action of $GL_N(\mathbb{Z})$ on X, has been studied for the last two centuries. When N is small it can be done explicitly and the stabilizers in $GL_N(\mathbb{Z})$ of points in X can be computed. This leads to explicit computation of the cohomology of $GL_N(\mathbb{Z})$ [39], and using the stability of K-theory, to the computation of $K_m(\mathbb{Z})$ for small values of m [23], [24], [40] (see also [45] for $K_3\left(\mathbb{Z}\left[\sqrt{-1}\right]\right)$). In [8], Borel was able to use Hodge theory on the (non compact!) manifold X/Γ to compute the rank of K-groups of algebraic integers (Chapter I, Theorem 1.2.). However reduction theory did not provide general results on the torsion in $K_m(\mathbb{Z})$.

III. The K-Theory of Complex Numbers With Finite Coefficients

We shall now forget about integers for a while, and sketch the proof of a result of Suslin [48] which computes the K-theory of complex numbers with finite coefficients. By Suslin's rigidity theorem [47] (see Gillet's lecture 7 in this volume) this is also the K-theory of $\bar{\mathbb{Q}}$ with finite coefficients. Hence, if the descent problem could be solved (i.e. if we could compute the K-theory of a field from the action of the Galois group on the K-theory of a Galois extension), these results of Suslin would lead to a computation of the K-theory of $\mathbb{Q}$, hence $\mathbb{Z}$. So the results in this chapter might turn out to be relevant in the future for the computation of the K-theory of $\mathbb{Z}$.

1. Statement of the Result

1.1. Let us come back to the notion of classifying space introduced in Chapter II, 1.1. Let G be a discrete group. The classifying space BG is the geometric realization of the simplicial set equal to G^p in degree p, with faces given by

$$d_i\left([g_1,\dots,g_p]\right) = \begin{cases} [g_2,\dots,g_p] & \text{if } i = 0 \\ [g_1,\dots,g_ig_{i+1},\dots,g_p] & \text{if } 1 \leq i \leq p-1 \\ [g_1,\dots,g_{p-1}] & \text{if } i = p \end{cases} ,$$

and degeneracies

$$s_i\left([g_1,\dots,g_p]\right) = [g_1,\dots,g_i,e,g_{i+1},\dots,g_p] ,$$

where e is the unit in G. This means that BG is the quotient of the disjoint union $\coprod_{p\geq 0} G^p \times \Delta(p)$, where $\Delta(p) = \{(x_0,\dots,x_p) \in \mathbb{R}^{p+1}, x_0 + \dots + x_p = 1\}$ is the standard simplex, by identifications provided by d_i and s_i. Here G^p has the discrete topology and $G^p \times \Delta(p)$ has the product topology.

Assume that G is a (locally compact) topological group. We can then define the classifying space BG^{top} of the topological group G by the same method, except that we endow G^p with the product topology instead of the discrete one. Clearly there is a map $BG \longrightarrow BG^{\text{top}}$.

1.2. Let $BGL_N(\mathbb{C})^{\text{top}}$ be the classifying space of the topological group $GL_N(\mathbb{C})$ and

$$BGL(\mathbb{C})^{\text{top}} = \varinjlim_N BGL_N(\mathbb{C})^{\text{top}} .$$

The map $BGL(\mathbb{C}) \longrightarrow BGL(\mathbb{C})^{\text{top}}$ factors through a map

$$\psi : BGL(\mathbb{C})^+ \longrightarrow BGL(\mathbb{C})^{\text{top}}$$

(to see it, one proves that $BGL(\mathbb{C})^{\text{top}}$ is a connected H-group and that any map from $BGL(A)$ to a connected H-group – for any ring A – factors through $BGL(A)^+$, by [27], Prop. 1.1.2.. See Loday's Lecture No. 5, Proposition 5.3).

The homotopy of $BGL(\mathbb{C})^{\text{top}}$ is known. If

$$U = \varinjlim_{N \geq 1} U_N$$

denotes the infinite unitary group, $BGL(\mathbb{C})^{\text{top}}$ is homotopic to BU^{top}. In order to use a more standard notation let us write $BU = BU^{\text{top}}$. The homotopy of BU is periodic and was computed by Bott [12]. One gets

$$\pi_m\left(BGL(\mathbb{C})^{\text{top}}\right) = \pi_m(BU) = \mathbb{Z}$$

when $m > 0$ is even, 0 when m is odd. If we take homotopy with coefficients we get $\pi_m\left(BGL(\mathbb{C})^{\text{top}}; \mathbb{Z}/n\right) = \mathbb{Z}/n$ if m is even, 0 if m is odd.

In this chapter we shall prove the following result of Suslin [48]:

Theorem 1.1. *For any $m \geq 0$ and $n > 0$ the map ψ induces an isomorphism*

$$K_m(\mathbb{C}; \mathbb{Z}/n) \xrightarrow{\sim} \pi_m(BU; \mathbb{Z}/n) \ .$$

As explained in Gillet's lecture No. 7, Theorem 1.1 allows one to prove the Lichtenbaum conjecture on the K-theory of any algebraically closed field of characteristic zero (i.e. the Corollary in 7.4, loc. cit.).

1.3. A first reduction of Theorem 1.1 is a "Hurewicz theorem modulo n", which shows that it is enough to prove that ψ induces an isomorphism in homology with coefficients $\mathbb{Z}/n$ for all $n > 0$. We refer the reader to [48] for details about this step.

1.4. A second reduction is to assume that n is prime. This is possible because, when n and n' are prime to each other $H_*(-; \mathbb{Z}/nn') = H_*(-; \mathbb{Z}/n) \oplus H_*(-; \mathbb{Z}/n')$. Furthermore if $n = p^k$ there is a natural long exact sequence:

$$\dots \longrightarrow H_*(-; \mathbb{Z}/p) \longrightarrow H_*(-; \mathbb{Z}/p^k) \longrightarrow H_*(-; \mathbb{Z}/p^{k-1})$$
$$\longrightarrow H_{*-1}(-; \mathbb{Z}/p) \longrightarrow \dots \ .$$

By induction on k one is reduced the case where $n = p$ is a prime number.

1.5. Since $BGL(\mathbb{C}) \longrightarrow BGL(\mathbb{C})^+$ induces an isomorphism on homology, we have to prove that the map

$$H_m\big(BGL(\mathbb{C}); \mathbb{Z}/n\big) \longrightarrow H_m\big(BGL(\mathbb{C})^{\text{top}}; \mathbb{Z}/n\big)$$

is an isomorphism. For this we can replace $GL(\mathbb{C})$ by $GL_N(\mathbb{C})$ for all N big enough with respect to m.

2. Description of a Fiber

2.1. Recall that any map f between spaces is homotopy equivalent to a fibration, the fiber of which is called the *homotopy fibre* of f. Let $G = GL_N(\mathbb{C})$, $N \geq 1$, and let ϕ be the fibre of the map $BG \longrightarrow BG^{\text{top}}$.

2.2. To prove that the map $BG \longrightarrow BG^{\text{top}}$ induces an isomorphism on $H_m(-; \mathbb{Z}/n)$ when $N \gg m$, it is enough to prove that, if $N \gg m$:

a) The map $H_m(BG, \mathbb{Z}/n) \longrightarrow H_m(BG^{\text{top}}, \mathbb{Z}/n)$ is surjective.
b) The map $H_m(\phi; \mathbb{Z}/n) \longrightarrow H_m(BG; \mathbb{Z}/n)$ is zero.

To see this, assume that a) and b) are true. By induction on $m' \ll N$ we prove that the reduced homology $\tilde{H}_{m'}(\phi; \mathbb{Z}/n)$ vanishes and that $H_{m'}(BG; \mathbb{Z}/n) \longrightarrow H_{m'}(BG^{\text{top}}; \mathbb{Z}/n)$ is an isomorphism. If it is true when $m' < m$, the spectral sequence

$$E^2_{pq} = H_p(BG^{\text{top}}, H_q(\phi; \mathbb{Z}/n)) \Rightarrow H_{p+q}(BG; \mathbb{Z}/n)$$

is such that $E^2_{pq} = 0$ when $0 < q < m$. This leads to an exact sequence:

$$H_{m+1}(BG; \mathbb{Z}/n) \xrightarrow{\alpha} H_{m+1}(BG^{\text{top}}; \mathbb{Z}/n) \longrightarrow H_m(\phi, \mathbb{Z}/n)$$
$$\xrightarrow{\beta} H_m(BG; \mathbb{Z}/n) \longrightarrow H_m(BG^{\text{top}}; \mathbb{Z}/n) \longrightarrow 0 \ .$$

Since α is surjective and β is zero (by a) and b)) we deduce that $H_m(\phi, \mathbb{Z}/n) = 0$ and $H_m(BG; \mathbb{Z}/n) = H_m(BG^{\text{top}}; \mathbb{Z}/n)$.

2.3. The proof of a) in 2.2 will be given in Chapter IV below (Prop. 1.1.). In this chapter we prove b). The first step consists in describing ϕ. Let $\varepsilon > 0$ be a positive real number, d an invariant metric on G, and G_ε the set of $g \in G$ such that $d(g, e) < \varepsilon$. If ε is small enough, for every $(g_1, \dots, g_p) \in G^p$, the open set

$$U(g_1, \dots, g_p) = G_\varepsilon \cap g_1 G_\varepsilon \cap g_1 g_2 G_\varepsilon \cap \dots \cap g_1 \dots g_p G_\varepsilon$$

is either empty or contractible. Let G^p_ε be the subset G^p made of those $(g_1, \dots, g_p)$ such that $U(g_1, \dots, g_p)$ is not empty. One checks easily that $d_i(G^p_\varepsilon) \subset G^{p-1}_\varepsilon$ and $s_i(G^p_\varepsilon) \subset G^{p+1}_\varepsilon$ for all i. Let BG_ε be the geometric realization of the simplicial set $(G^p_\varepsilon, d_i, s_i)$ and $BG_\varepsilon \longrightarrow BG$ the inclusion. One can prove that ϕ is homotopic to BG_ε, i.e.

Proposition 2.3. *When ε is small enough, the sequence of maps*

$$BG_\varepsilon \longrightarrow BG \longrightarrow BG^{\text{top}}$$

is homotopy equivalent to a fibration.

Proof. cf. [48], Prop. 4.1.

2.4. From the stability results in homology we know that the map

$$H_m(BGL_N(\mathbb{C}), \mathbb{Z}/n) \longrightarrow H_m(BGL(\mathbb{C}), \mathbb{Z}/n)$$

is an isomorphism when $N \gg m$ (see [46] and Chapter II, Theorem 2.2). Hence, to prove b) in 2.2., we need, by Proposition 2.3, the following result: given N and m positive integers, there exists $\varepsilon > 0$ such that the composite map

$$H_m(BGL_N(\mathbb{C})_\varepsilon; \mathbb{Z}/n) \longrightarrow H_m(BGL_N(\mathbb{C}), \mathbb{Z}/n) \longrightarrow H_m(BGL(\mathbb{C}), \mathbb{Z}/n)$$

is zero. Suslin's proof of this fact will be described in the next two paragraphs.

3. K-Theory of Hensel Rings

3.1. Let X be a smooth irreducible algebraic complex variety, $d = \dim(X)$ its dimension, and $x \in X(\mathbb{C})$ a point on X. Let $\mathcal{O}_x$ be the local ring of rational functions on X which are defined at x, $\mathcal{M}_x \subset \mathcal{O}_x$ its maximal ideal (functions which vanish at x), and

$$\hat{\mathcal{O}}_x = \varprojlim_p \left(\mathcal{O}_x / (\mathcal{M}_x)^p \right)$$

the completion of $\mathcal{O}_x$. The ring $\hat{\mathcal{O}}_x$ is isomorphic to the ring $\mathbb{C}[[t_1, \ldots, t_d]]$ of power series in d variables, where $t_1, \ldots, t_d$ are local parameters at x. The henselization of $\mathcal{O}_x$ is, by definition, the integral closure $\mathcal{O}_x^h$ of $\mathcal{O}_x$ inside $\hat{\mathcal{O}}_x$, i.e. the set of elements α in $\hat{\mathcal{O}}_x$ such that there exist $a_1, \ldots, a_p$ in $\mathcal{O}_x$ with

$$\alpha^p + a_1 \alpha^{p-1} + \ldots + a_p = 0 \ . \tag{3.1.1}$$

Let $\mathcal{M}_x^h = \mathcal{M}_x \cdot \mathcal{O}_x^h$ be the maximal ideal in $\mathcal{O}_x^h$. We have $\mathcal{O}_x^h / \mathcal{M}_x^h \simeq \mathcal{O}_x / \mathcal{M}_x \simeq \mathbb{C}$, where the last isomorphism is given by evaluation at the point x.

A *basic remark* is the following: any power series α in $\mathcal{O}_x^h \subset \mathbb{C}[[t_1, \ldots, t_d]]$ has a strictly positive convergence radius (this follows from (3.1.1.)).

Theorem 3.2. [20],[19]. *For any* $m > 0$ *the projection*

$$\mathcal{O}_x^h \longrightarrow \mathcal{O}_x^h / \mathcal{M}_x^h \simeq \mathbb{C}$$

induces an isomorphism

$$K_m \left(\mathcal{O}_x^h ; \mathbb{Z}/n \right) \longrightarrow K_m(\mathbb{C}; \mathbb{Z}/n) \ .$$

We shall not give the proof of this result. It is an extension of the rigidity theorem of Suslin for algebraically closed fields (cf. Gillet's lecture No. 7, 7.4., Remark i)). A more general result holds for any henselian pair (A, I) instead of $(\mathcal{O}_x^h, \mathcal{M}_x^h)$ [19].

3.3. Theorem 3.2. implies that the map

$$H_m \left(GL \left(\mathcal{O}_x^h \right) ; \mathbb{Z}/n \right) \longrightarrow H_m \left(GL(\mathbb{C}); \mathbb{Z}/n \right)$$

is an isomorphism for all $m \geq 0$. Let $GL \left(\mathcal{O}_x^h, \mathcal{M}_x^h \right)$ be the kernel of the map $GL \left(\mathcal{O}_x^h \right) \longrightarrow GL(\mathbb{C})$. Using the Hochschild-Serre spectral sequence of the extension of groups

$$1 \longrightarrow GL \left(\mathcal{O}_x^h, \mathcal{M}_x^h \right) \longrightarrow GL \left(\mathcal{O}_x^h \right) \longrightarrow GL(\mathbb{C}) \longrightarrow 1$$

we get $H_m \left(GL \left(\mathcal{O}_x^h, \mathcal{M}_x^h \right) ; \mathbb{Z}/n \right) = 0$ for all $m \geq 1$.

4. The Universal Chain Construction

4.1. Fix integers $M \geq 1$ and $m \geq 1$. Consider the complex affine variety

$$GL_M^m = GL_M \times \ldots \times GL_M \qquad (m \text{ copies}) \ .$$

Let A_m be the ring of regular functions on GL_M^m, $\mathcal{O}_m$ the henselization of the local ring of GL_M^m at the origin $(e, \ldots, e) \in GL_M(\mathbb{C})^m$ (where $e \in GL_M(\mathbb{C})$ is the identity), and $\mathcal{M}_m$ the maximal ideal in $\mathcal{O}_m$.

There is a canonical map (obtained by restriction) $A_m \longrightarrow \mathcal{O}_m$. Furthermore, if $j = 1, \ldots, m$, the projection $\pi_j : GL_M^m \longrightarrow GL_M$ on the j-th factor is given by a ring homomorphism $\pi_j^* : A_1 \longrightarrow A_m$.

The composite map

$$A_1 \xrightarrow{\pi_j^*} A_m \longrightarrow \mathcal{O}_m$$

defines a point in $GL_M(\mathcal{O}_m)$. Let α_j be its image in $GL(\mathcal{O}_m)$. Since the composite map

$$A_1 \xrightarrow{\pi_j^*} A_m \longrightarrow \mathcal{O}_m \longrightarrow \mathbb{C}$$

is the evaluation of functions in A_1 at the origin e, we obtain that α_j lies in the kernel $GL(\mathcal{O}_m; \mathcal{M}_m)$ of the projection map

$$GL(\mathcal{O}_m) \longrightarrow GL(\mathcal{O}_m / \mathcal{M}_m) \ .$$

Let $u_m = [\alpha_1, \ldots, \alpha_m] \in GL(\mathcal{O}_m; \mathcal{M}_m)^m$.

4.2. If $0 \leq j \leq m$ define a morphism of algebraic varieties

$$P_j^m : GL_M^m \longrightarrow GL_M^{m-1}$$

by the formula

$$P_j^m(g_1, \ldots, g_m) = \begin{cases} (g_2, \ldots, g_m) & \text{if } j = 0 \\ (g_1, \ldots, g_j g_{j+1}, \ldots, g_m) & \text{if } 1 \leq j \leq m-1 \\ (g_1, \ldots, g_{m-1}) & \text{if } j = m \end{cases} \ .$$

Since $P_j^m(e, \ldots, e) = (e, \ldots, e)$ we get a map of local rings $\left(P_j^m\right)^* : \mathcal{O}_{m-1} \longrightarrow \mathcal{O}_m$.

On the other hand, given a discrete group G, let $C_m(G) = (\mathbb{Z}/n)[G^m]$ be the free $\mathbb{Z}/n$-module generated by G^m and $d : C_m(G) \longrightarrow C_{m-1}(G)$ the standard inhomogeneous boundary map (see 1.1.):

$$d(g_1, \ldots, g_m) = (g_2, \ldots, g_m) + \sum_{j=1}^{m-1} (-1)^j (g_1, \ldots, g_j g_{j+1}, \ldots, g_m)$$

$$+ (-1)^m (g_1, \ldots, g_{m-1}) \ .$$

4.3. Let $u_m = [\alpha_1, \ldots, \alpha_m] \in C_m\left(GL(\mathcal{O}_m, \mathcal{M}_m)\right)$ be the chain defined in 4.1.

Proposition 4.3. *For all $m \geq 0$ there exists a chain $c_m \in C_{m+1}\left(GL(\mathcal{O}_n, \mathcal{M}_m)\right)$ such that, if $m \geq 1$,*

$$d(c_m) = u_m - \sum_{j=0}^{m} (-1)^j \left(P_j^m\right)^* (c_{m-1}) \ . \tag{4.3.1}$$

Proof. We construct c_m be induction on m. Take $c_0 = 0$. Assume $c_0, \ldots, c_{m-1}$ have been constructed. Let

$$x_m = u_m - \sum_{j=0}^{m} (-1)^j \left(P_j^m\right)^* (c_{m-1}) \ .$$

When $m = 1$ we get
$$d(x_1) = d(u_1) = d([\alpha_1]) = 0 \ .$$

When $m \geq 2$ we use the formula (4.3.1) for $d(c_{m-1})$ to get

$$d(x_m) = d(u_m)$$
$$- \sum_{j=0}^{m} (-1)^j \left(P_j^m\right)^* \left[u_{m-1} - \sum_{k=0}^{m-1} (-1)^k \left(P_k^{m-1}\right)^* (c_{m-2}) \right] \ .$$

But
$$d(u_m) = d\left([\alpha_1, \ldots, \alpha_m]\right)$$
$$= [\alpha_2, \ldots, \alpha_m] + \sum_{j=1}^{m-1} (-1)^j [\alpha_1, \ldots, \alpha_j \alpha_{j+1}, \ldots, \alpha_m]$$
$$+ (-1)^m [\alpha_1, \ldots, \alpha_{m-1}]$$
$$= \sum_{j=0}^{m} (-1)^j \left(P_j^m\right)^* (u_{m-1}) \ .$$

Furthermore (since $d \circ d = 0$) we have

$$\sum_{j=0}^{m} \sum_{k=0}^{m-1} (-1)^{j+k} \left(P_j^m\right)^* \left(P_k^{m-1}\right)^* = 0 \ .$$

Therefore $d(x_m) = 0$.

Since $H_m\left(GL(\mathcal{O}_m, \mathcal{M}_m); \mathbb{Z}/n\right) = 0$ (see 3.3.), x_m is a boundary, i.e. there exists c_m in $C_{m+1}\left(GL(\mathcal{O}_m, \mathcal{M}_m)\right)$ such that $d(c_m) = x_m$. $\square$

4.4. Let g be a matrix in $GL(\mathcal{O}_m, \mathcal{M}_m)$. The coefficients of $g - 1$ are in $\mathcal{M}_m$ and almost all of them are zero. Using the basic remark of 3.1., we see that, if $\eta > 0$ is small enough, these coefficients define maps from $GL_M(\mathbb{C})_\eta^m$ to $\mathbb{C}$. Therefore g gives a map

$$GL_M(\mathbb{C})_\eta^m \longrightarrow GL(\mathbb{C}) \ .$$

Given $(g_1,\ldots,g_p) \in GL(\mathcal{O}_m,\mathcal{M}_m)^p$ we get, for $\eta > 0$ small enough, a map

$$GL_M(\mathbb{C})_\eta^m \longrightarrow GL(\mathbb{C})^p \ .$$

By linearity, any chain c in $C_p(GL(\mathcal{O}_m,\mathcal{M}_m))$ defines a map of $(\mathbb{Z}/n)$-modules

$$\tau(c) : C_m(GL_N(\mathbb{C})_\eta) \longrightarrow C_p(GL(\mathbb{C}))$$

(for all $\eta > 0$ small enough).

4.5. Using the definitions above we see that $\tau(u_m) : C_m\big(BGL_N(\mathbb{C})_\eta\big) \longrightarrow C_m\big(GL(\mathbb{C})\big)$ is the map induced by the inclusion $GL_N(\mathbb{C})_\eta \longrightarrow GL(\mathbb{C})$. Furthermore, given c in $C_p(GL(\mathcal{O}_m,\mathcal{M}_m))$ we get

$$\tau\big(d(c)\big) = d \circ \tau(c)$$

and

$$\tau\left(\sum_{j=0}^m (-1)^j \left(P_j^m\right)^* (c)\right) = \tau(c) \circ d \ .$$

4.6. Let $\varepsilon > 0$ be small enough for $\tau(c_m)$, $\tau(c_{m-1})$ and $\tau(u_m)$ to be defined in an ε neighbourhood of the origin. Using 4.5. and Proposition 4.3. we get $\tau \circ d(c_m) = \tau(u_m) - \tau(c_{m-1}) \circ d$. If $x \in C_m(BGL_N(\mathbb{C})_\varepsilon)$ is a cycle we deduce that, since $d(x) = 0$,

$$\tau(u_m)(x) = d\left(\tau(c_m)(x)\right) \ ,$$

i.e. the image of x in $C_m\big(GL(\mathbb{C})\big)$ is a boundary. In other words, the map

$$H_m\big(BGL_N(\mathbb{C})_\varepsilon; \mathbb{Z}/n\big) \longrightarrow H_m\big(BGL(\mathbb{C}); \mathbb{Z}/n\big)$$

is zero, as was to be shown (cf. 2.4.).

IV. Chern Classes of Discrete Groups

In this chapter we complete the computation of the K-theory of complex numbers with finite coefficients (Chapter III, Thm. 1.1.) by introducing the Chern classes. We show that these give also interesting torsion classes in the integral cohomology of $GL_N(\mathbb{Z})$.

1. Chern Classes for the Discrete Group $GL_N(\mathbb{C})$

1.1. To end the proof of Theorem 1.1 in Chapter III it suffices to show (Chapter III, 2.2.a)):

Proposition 1.1. *Let p be a prime integer and $N \geq 1$. The map of classifying spaces $BGL_N(\mathbb{C}) \longrightarrow BGL_N(\mathbb{C})^{\text{top}}$ induces a surjection on the homology with $\mathbb{Z}/p$ coefficients*

$$H_m\big(BGL_N(\mathbb{C}), \mathbb{Z}/p\big) \twoheadrightarrow H_m\big(BGL_n(\mathbb{C})^{\text{top}}, \mathbb{Z}/p\big) \ ,$$

for every $m \geq 1$.

1.2. Let $\mu_p \subset \mathbb{C}^*$ be the group of p-th roots of unity. The direct product μ_p^N of N copies of μ_p can be imbedded diagonally into $GL_N(\mathbb{C})$. To prove Proposition 1.1. we will show that the composite map

$$H_m\big(\mu_p^N, \mathbb{Z}/p\big) \longrightarrow H_m\big(GL_N(\mathbb{C}), \mathbb{Z}/p\big) \longrightarrow H_m\big(BGL_N(\mathbb{C})^{\text{top}}, \mathbb{Z}/p\big)$$

is surjective.

Since both μ_p^N and $BGL_N(\mathbb{C})^{\text{top}}$ have finitely generated homology and p is a prime, their cohomology is dual to their homology and we have to show that

$$H^m\big(BGL_N(\mathbb{C})^{\text{top}}; \mathbb{Z}/p\big) \longrightarrow H^m\big(\mu_p^N, \mathbb{Z}/n\big)$$

is injective. The cohomology of $BGL_N(\mathbb{C})^{\text{top}}$ with integral coefficients is known to be the polynomial algebra $\mathbb{Z}[c_1, \ldots, c_N]$, where $c_i \in H^{2i}\big(BGL_N(\mathbb{C})^{\text{top}}, \mathbb{Z}\big)$ is the i-th universal Chern class [31]. Therefore $H^*\big(BGL_N(\mathbb{C})^{\text{top}}, \mathbb{Z}/p\big)$ is isomorphic to $\mathbb{Z}/p[\overline{c_1}, \ldots \overline{c_N}]$, where $\overline{c_i}$ is the reduction of c_i modulo p.

The pull-back of the universal bundle on $BGL_N(\mathbb{C})^{\text{top}}$ to $B\mu_p^N$ is a direct sum of line bundles, since the restriction of the natural representation of $GL_N(\mathbb{C})$ to μ_p^n splits as a direct sum of characters $\chi_1, \ldots, \chi_N$. Hence the pull-back of $\overline{c_i}$ to $H^{2i}\big(B\mu_p^N, \mathbb{Z}/p\big)$ is the reduction modulo p of the elementary symmetric function $\sigma_i\big(c_1(\chi_1), \ldots, c_1(\chi_N)\big) \in H^{2i}\big(B\mu_p^N, \mathbb{Z}\big)$, where $c_1(\chi_\alpha)$ is the first Chern class of the line bundle on $B\mu_p^N$ attached to the character χ_α given by the α-th projection:

$$\mu_p^N \xrightarrow{\pi_\alpha} \mu_p \longrightarrow \mathbb{C}^* \ .$$

By the Kunneth formula

$$H^*\big(\mu_p^N, \mathbb{Z}/p\big) = \overset{N}{\underset{\alpha=1}{\otimes}} H^*(\mu_p, \mathbb{Z}/p) \ .$$

Therefore we just need to show that, for every $k \geq 0$, $\overline{c_1}(\chi)^k$ is non-zero in $H^{2k}(\mu_p, \mathbb{Z}/p)$, where $\chi : \mu_p \longrightarrow \mathbb{C}^*$ is the inclusion.

1.3. More generally we shall prove that $\overline{c_1}(\chi)^k$ is non zero in $H^{2k}(\mu_n, \mathbb{Z}/n)$ for an arbitrary integer $n \geq 1$ (and $\chi : \mu_n \longrightarrow \mathbb{C}^*$ the inclusion). The first Chern class of a line bundle L on a topological space X is the image of the class of L

in $H^1(X, \mathbb{C}^*)$ by the Bockstein map $\beta : H^1(X, \mathbb{C}^*) \longrightarrow H^2(X, \mathbb{Z})$ coming from the exact sequence of coefficients

$$0 \longrightarrow \mathbb{Z} \longrightarrow \mathbb{C} \stackrel{\exp(2\pi i \cdot)}{\longrightarrow} \mathbb{C}^* \longrightarrow 0 .$$

Therefore $c_1(\chi) \in H^2(\mu_n, \mathbb{Z})$ is the image of the character $\chi \in H^1(\mu_n, \mathbb{C}^*) = Hom(\mu_n, \mathbb{C}^*)$ by the Bockstein map β. Since μ_n is a finite group we have $H^k(\mu_n, \mathbb{C}) = 0$ when $k > 0$, and β is an isomorphism. Therefore $c_1(\chi) = \beta(\chi)$ is a generator of the cyclic group $H^2(\mu_n, \mathbb{Z}) \simeq Hom(\mu_n, \mathbb{C}^*) = \mathbb{Z}/n$.

The cyclic group μ_n is known to have periodic cohomology: the cup-product by any generator of $H^2(\mu_n, \mathbb{Z})$ induces an isomorphism from $H^k(\mu_n, \mathbb{Z})$ to $H^{k+2}(\mu_n, \mathbb{Z})$, $k \geq 0$ [14]. Therefore $H^k(\mu_n, \mathbb{Z}) = 0$ when n is odd (since $H^1(\mu_n, \mathbb{Z}) = Hom(\mu_n, \mathbb{Z}) = 0$) and for every k, $c_1(\chi)^k$ is a generator in $H^{2k}(\mu_n, \mathbb{Z})$. From the exact sequence of coefficients

$$0 \longrightarrow \mathbb{Z} \stackrel{n}{\longrightarrow} \mathbb{Z} \longrightarrow \mathbb{Z}/n \longrightarrow 0$$

we deduce (since $nH^k(\mu_n, \mathbb{Z}) = 0$) that the map $H^{2k}(\mu_n, \mathbb{Z}) \longrightarrow H^{2k}(\mu_n, \mathbb{Z}/n)$ is an isomorphism. Hence $\overline{c_1}(\chi)^k \neq 0$, as was to be shown.

2. Chern Classes for the Discrete Group $GL_N(\mathbb{Z})$

2.1. Let $N \geq 1$ and denote by

$$\omega : BGL_N(\mathbb{Z}) \longrightarrow BGL_N(\mathbb{C})^{\text{top}}$$

the composite of $BGL_N(\mathbb{Z}) \longrightarrow BGL_N(\mathbb{C})$ (induced by the inclusion $\mathbb{Z} \subset \mathbb{C}$) with $BGL_N(\mathbb{C}) \longrightarrow BGL_N(\mathbb{C})^{\text{top}}$. We can consider the Chern classes

$$\omega^*(c_i) \in H^{2i}(GL_N(\mathbb{Z}), \mathbb{Z}), \quad i > 0 .$$

Since ω factors through $BGL_N(\mathbb{C})$ we know from Gillet's lecture No.6, 6.3., that $\omega^*(c_i)$ is a torsion element. We shall study its order. Let b_i be the i-th Bernoulli number (Chapter I 2.1.) and $\text{den}(b_i/i)$ the denominator of b_i/i.

Theorem 2.1. *The order of $\omega^*(c_i) \in H^{2i}(GL_N(\mathbb{Z}), \mathbb{Z})$ is equal to $2\text{den}(b_i/i)$ when 4 does not divide i. When 4 divides i, it is either $\text{den}(b_i/i)$ or $2\text{den}(b_i/i)$.*

2.2. Here we shall only prove that the order of $\omega^*(c_i)$ is divisible by the odd part of $\text{den}(b_i/i)$ [18], [42]. For this we restrict $\omega^*(c_i)$ to certain finite subgroups of $GL_N(\mathbb{Z})$. Let us fix an integer $n \geq 1$ and consider the subfield $\mathbb{Q}(\mu_n) \subset \bar{\mathbb{Q}}$ of the algebraic closure of $\mathbb{Q}$ generated by the n-th roots of unity. The ring of integers $\mathbb{Z}[\mu_n]$ in $\mathbb{Q}(\mu_n)$ is a free $\mathbb{Z}$-module of rank $r = (\mathbb{Q}(\mu_n) : \mathbb{Q})$. If we choose a basis for this module, the action of μ_n upon it will give an injective morphism $\mu_n \longrightarrow GL_r(\mathbb{Z})$. Let $\Gamma_n \subset GL_r(\mathbb{Z})$ be its image. We shall restrict to Γ_n the classes $\omega^*(c_i) \in H^{2i}(GL_N(\mathbb{Z}), \mathbb{Z})$.

Lemma 2.2. *Let p be an odd prime, $\nu \geq 1$ and $n = p^\nu$. When $i = p^\nu - p^{\nu-1}$ the restriction of $\omega^*(c_i)$ to Γ_n has order n.*

Proof of Lemma 2.2. Let $g \in \Gamma_n$ be a generator. Since $g^{p^\nu} = 1$ and $g^{p^{\nu-1}} \neq 1$, the characteristic polynomial of the matrix g is the cyclotomic polynomial $\left(X^{p^\nu} - 1\right) / \left(X^{p^{\nu-1}} - 1\right)$. Since the roots of this polynomial in $\mathbb{C}$ are all distinct (these are the primitive p^ν-th roots of unity) g can be diagonalized in $GL_r(\mathbb{C})$. In other words the complex representation $\rho : \Gamma_n \longrightarrow GL_r(\mathbb{C})$ is the direct sum of r distinct characters of Γ_n, obtained by sending g to a primitive root of unity. Let us fix an imbedding $\mathbb{Q}(\mu_n) \subset \mathbb{C}$, hence a character $\chi : \mu_n = \Gamma_n \longrightarrow \mathbb{C}^*$.

Now ρ is the direct sum

$$\rho = \bigoplus_{\sigma \in G} \chi \circ \sigma \ ,$$

where $\chi \circ \sigma$ is the translate of χ by $\sigma \in G = \mathrm{Gal}\big(\mathbb{Q}(\mu_n)/\mathbb{Q}\big)$. The total Chern class $\sum_{i \geq 0} \omega^*(c_i)$ (with $c_0 = 1$), restricted to Γ_n, maps to the total Chern class of ρ:

$$c(\rho) = \prod_{\sigma \in G} (1 + c_1(\chi \circ \sigma)) \ . \tag{2.2.1.}$$

Let us compute $c_1(\chi \circ \sigma) \in H^2(\Gamma_n, \mathbb{Z})$. Let us define $\varepsilon(\sigma) \in (\mathbb{Z}/n)^*$ by the equations

$$\sigma(\zeta) = \zeta^{\varepsilon(\sigma)} \qquad \text{for all} \quad \zeta \in \mu_n \ . \tag{2.2.2.}$$

In this way we get an isomorphism

$$\varepsilon : \mathrm{Gal}\big(\mathbb{Q}(\mu_n)/\mathbb{Q}\big) \longrightarrow (\mathbb{Z}/n)^*$$

called the cyclotomic character. Given $\zeta \in \mu_n \simeq \Gamma_n$ we have

$$\chi \circ \sigma(\zeta) = \chi\left(\zeta^{\varepsilon(\sigma)}\right) = \chi(\zeta)^{\varepsilon(\sigma)} \ ,$$

hence $\chi \circ \sigma = \varepsilon(\sigma)\chi$ and $c_1(\chi \circ \sigma) = \beta(\chi \circ \sigma) = \varepsilon(\sigma)c_1(\chi)$.

Using (2.2.1.) we get

$$c(\rho) = \prod_{\sigma \in G} (1 + \varepsilon(\sigma)c_1(\chi)) \ . \tag{2.2.3.}$$

Since $p \neq 2$ the group $(\mathbb{Z}/n)^*$ is cyclic of order $r = p^\nu - p^{\nu-1}$. Since $\varepsilon(\sigma)$, $\sigma \in G$, runs over all elements in $(\mathbb{Z}/n)^*$, we get

$$\prod_{\sigma} (1 + \varepsilon(\sigma)c_1(\chi)) = 1 + \ldots - c_1(\chi)^{p^\nu - p^{\nu-1}} \ . \tag{2.2.4.}$$

We saw in 1.3 that, for every $k > 0$, the element $c_1(\chi)^k \in H^{2k}(\mu_n, \mathbb{Z}) \simeq H^{2k}(\mu_n, \mathbb{Z}/n)$ has order n. Therefore, by (2.2.3.) and (2.2.4.), when $i = p^\nu - p^{\nu-1}$ the element $c_i(\rho)$ has order $n = p^\nu$, as was to be shown.

2.3. Let $i > 0$ be an arbitrary integer, p an odd prime number such that $p - 1$ divides i, and $\nu > 0$ the integer such that

$$i = j(p^\nu - p^{\nu-1})$$

with j prime to p. Given $N \geq i$ the direct sum map

$$GL_r(\mathbb{Z})^j \longrightarrow GL_N(\mathbb{Z})$$

(with $r = p^\nu - p^{\nu-1}$) gives an imbedding of Γ_n^j into $GL_N(\mathbb{Z})$ (with $n = p^\nu$). Using the addition formula for Chern classes, the Kunneth formula and Lemma 2.2., we see that the restriction of $\omega^*(c_i)$ to Γ_n^j has order p^ν.

A theorem of von Staudt ([11], 5.8., Theorem 4, p. 384) asserts that a prime p divides $\text{den}(b_i/i)$ if and only if $p-1$ divides i, and that p^ν is the exact power of p dividing $\text{den}(b_i/i)$ when $i = j(p^\nu - p^{\nu-1})$ with j prime to p. From this we conclude that the odd factor of $\text{den}(b_i/i)$ divides the order of $\omega^*(c_i)$. When $p = 2$, the same argument applies, except that $(\mathbb{Z}/2^\nu)^*$ is the product of $\mathbb{Z}/2$ by a cyclic group. One gets [18], [42] that $\text{den}(b_i/i)$ divides the order of $\omega^*(c_i)$.

2.4. In [21], Grothendieck used étale cohomology to prove that $2\text{den}(b_i/i)\omega^*(c_i) = 0$ (see also Chapter V).

When i is not divisible by 4 it can be proved that $\omega^*(c_i)$ has order exactly $2\text{den}(b_i/i)$ [17], [2]. When i is divisible by 4, I do not know whether $\omega^*(c_i)$ has order $\text{den}(b_i/i)$ or $2\text{den}(b_i/i)$.

V. Étale Cohomology

In Chapter IV we saw that denominators of Bernoulli numbers are related to the order of Chern classes of $GL_N(\mathbb{Z})$. We shall now consider the K-theory of $\mathbb{Z}$ and see that both denominators and numerators of Bernoulli numbers give lower bounds for the torsion in $K_m(\mathbb{Z})$.

1. Denominators

1.1. The maps

$$\omega : BGL_N(\mathbb{Z}) \longrightarrow BGL_N(\mathbb{C})^{\text{top}}, \quad N \geq 1 ,$$

induced by the inclusion $GL_N(\mathbb{Z}) \subset GL_N(\mathbb{C})$ are compatible with the stabilization map $GL_N \longrightarrow GL_{N+1}$. Hence we get a map $BGL(\mathbb{Z}) \longrightarrow BGL(\mathbb{C})^{\text{top}}$ by taking the inductive limit on N. Since $BGL(\mathbb{C})^{\text{top}}$ is a connected H-group this map factors through

$$\varphi : BGL(\mathbb{Z})^+ \longrightarrow BGL(\mathbb{C})^{\text{top}} .$$

The map φ is the composite of $BGL(\mathbb{Z})^+ \longrightarrow BGL(\mathbb{C})^+$ with the map ψ considered in III, 1.2. For every prime p and every integer $\nu \geq 1$, it induces a morphism

$$\varphi_* : K_m(\mathbb{Z}, \mathbb{Z}/p^\nu) \longrightarrow \pi_m(BGL(\mathbb{C})^{\text{top}}, \mathbb{Z}/p^\nu)$$

on homotopy groups with coefficients (see Gillet's lecture No.7 for the definitions). When $m = 2i$ we know that $\pi_m\big(BGL(\mathbb{C})^{\text{top}}, \mathbb{Z}/p^\nu\big) = \pi_m(BU, \mathbb{Z}/p^\nu) = \mathbb{Z}/p^\nu$, i.e. we get

$$\varphi_* : K_{2i}(\mathbb{Z}; \mathbb{Z}/p^\nu) \longrightarrow \mathbb{Z}/p^\nu \ .$$

Theorem 1.1. [13]. *Assume* $i = j\left(p^\nu - p^{\nu-1}\right)$ *with* j *prime to* p. *Then the map*

$$\varphi_* : K_{2i}(\mathbb{Z}, \mathbb{Z}/p^\nu) \longrightarrow \mathbb{Z}/p^\nu$$

is surjective.

1.2. We shall indicate the proof of Theorem 1.1. only when $\nu = 1$ and $p \neq 2$ (we refer to [13], Theorem 3.1. for the general case). First we choose a p-th root of unity $\zeta \in \mathbb{C}^*$, $\zeta \neq 1$. Let us consider the ring $A = \mathbb{Z}[\zeta]$, and the map

$$\varphi_A : BGL(A)^+ \longrightarrow BGL(\mathbb{C})^{\text{top}}$$

induced by the inclusion of A into $\mathbb{C}$. We shall see that the induced map

$$\varphi_{A*} : K_2(A; \mathbb{Z}/p) \longrightarrow \mathbb{Z}/p$$

is surjective . Consider the exact sequence

$$0 \longrightarrow K_m(A)/p \longrightarrow K_m(A; \mathbb{Z}/p) \longrightarrow K_{m-1}(A)[p] \longrightarrow 0 \ .$$

Since $K_1(A) = A^*$ ([32], Cor. 16.3.) we get

$$K_1(A)[p] = \mu_p \ ,$$

and there is an exact sequence

$$0 \longrightarrow K_2(A)/p \longrightarrow K_2(A; \mathbb{Z}/p) \longrightarrow \mu_p \longrightarrow 0 \ .$$

Let us define an element $\beta \in K_2(A; \mathbb{Z}/p)$ which maps to $\zeta \in \mu_p$. The composite inclusion of groups $i : \mu_p \longrightarrow A^* \simeq GL_1(A) \longrightarrow GL(A)$ induces on the homotopy of classifying spaces

$$i_* : \pi_2(B\mu_p; \mathbb{Z}/p) \longrightarrow K_2(A; \mathbb{Z}/p) \ .$$

Since $\pi_2(B\mu_p) = 0$ we know that

$$\pi_2(B\mu_p, \mathbb{Z}/p) \simeq \pi_1(B\mu_p)[p] \simeq \mu_p \ .$$

Hence ζ defines an element in $\pi_2(B\mu_p, \mathbb{Z}/p)$ and we can consider $\beta = i_*(\zeta) \in K_2(A; \mathbb{Z}/p)$.

Its image in $\mu_p = \pi_1(B\mu_p)[p] \simeq K_1(A)[p]$ is precisely ζ.

We want to prove that $\varphi_{A*}(\beta) \in \pi_2(BU; \mathbb{Z}/p)$ is non zero. Let us consider the representation

$$\chi = \varphi_A \circ i : \mu_p \longrightarrow GL(\mathbb{C})$$

and its first Chern class $\chi^*(\overline{c_1}) \in H^2(B\mu_p; \mathbb{Z}/p)$. Let $H : \pi_m(\cdot; \mathbb{Z}/p) \longrightarrow H_m(\cdot; \mathbb{Z}/p)$ be the Hurewicz map with $\mathbb{Z}/p$ coefficients. From the commutative diagram

$$\begin{array}{ccc} \pi_2(B\mu_p; \mathbb{Z}/p) & \xrightarrow{\sim} & \pi_1(B\mu_p) \\[1em] \downarrow H & & \downarrow \imath \\[1em] H_2(B\mu_p; \mathbb{Z}/p) & \longrightarrow & H_1(B\mu_p, \mathbb{Z}) \end{array}$$

we see that $H(\zeta) \in H_2(B\mu_p; \mathbb{Z}/p) = \mathbb{Z}/p$ is not zero. Since $\chi^*(\overline{c_1}) \in H^2(B\mu_p; \mathbb{Z}/p)$ is also not zero (IV. 1.3.) we get $\chi^*(\overline{c_1})(H(\zeta)) \neq 0$. But

$$\chi^*(\overline{c_1})(H(\zeta)) = \overline{c_1}\Big(H\big(\chi_*(\zeta)\big)\Big) = \overline{c_1}\Big(H\big(\varphi_{A*}(\beta)\big)\Big) \ ,$$

hence $\varphi_{A*}(\beta) \neq 0$ (in fact $\overline{c_1}\Big(H\big(\varphi_{A*}(\beta)\big)\Big) = 1$).

The element $\beta \in K_2(A; \mathbb{Z}/p)$ is called the "Bott element" because its image under φ_{A*} is the reduction modulo p of the generator $B \in \pi_2(BU) = \mathbb{Z}$ defined by Bott [12].

1.3. The tensor product of A-modules can be used to define a cup-product

$$K_m(A) \times K_n(A) \longrightarrow K_{m+n}(A)$$

(cf. Loday's lecture No. 7). The same is true for the K-theory with coefficients [13]. Hence we can consider powers of the Bott element, $\beta^i \in K_{2i}(A; \mathbb{Z}/p)$. The map φ_{A*} is multiplicative, hence $\varphi_{A*}(\beta^i) = \varphi_{A*}(\beta)^i$ is the reduction modulo p of the generator B^i of $\pi_{2i}(BU) = \mathbb{Z}$. Hence $\varphi_{A*}(\beta^i) \neq 0$.

To get an element in $K_{2i}(\mathbb{Z}; \mathbb{Z}/p)$ we use the transfer map. Namely, if we choose a basis of A over $\mathbb{Z}$, any matrix with coefficients in A can be viewed as a matrix over $\mathbb{Z}$, and we get inclusions

$$GL_N(A) \subset GL_{Nr}(\mathbb{Z}), \quad N \geq 1 \ ,$$

with $r = [A : \mathbb{Z}]$. These induce a map

$$BGL(A)^+ \longrightarrow BGL(\mathbb{Z})^+$$

and, on the homotopy groups with coefficients, a transfer $\mathrm{tr} : K_m(A; \mathbb{Z}/p) \longrightarrow K_m(\mathbb{Z}; \mathbb{Z}/p)$, $m > 0$.

On the other hand the inclusion $\alpha : \mathbb{Z} \longrightarrow A$ induces $\alpha_* : K_m(\mathbb{Z}; \mathbb{Z}/p) \longrightarrow K_m(A; \mathbb{Z}/p)$.

One can show that, for every $x \in K_m(A; \mathbb{Z}/p)$,

$$\alpha_* \mathrm{tr}(x) = \sum_{\sigma \in G} \sigma_*(x) \ ,$$

where $G = \mathrm{Gal}\big(\mathbb{Q}(\mu_p)/\mathbb{Q}\big)$ ([13], Lemmas 2.7 and 2.9).

In particular

$$\alpha_* \mathrm{tr}(\beta^i) = \sum_{\sigma \in G} \sigma_*(\beta^i) = \sum_{\sigma \in G} \sigma_*(\beta)^i \ .$$

Since the map $i : \mu_p \longrightarrow GL(A)$ is Galois equivariant we get

$$\sigma_*(\beta) = \sigma_*\big(i_*(\zeta)\big) = i_*\sigma_*(\zeta) = \varepsilon(\sigma)\beta \ ,$$

where ε is the cyclotomic character (Chapter IV,(2.2.2.)).

We conclude that

$$\alpha_* \mathrm{tr}(\beta^i) = \left(\sum_{\sigma \in G} \varepsilon(\sigma)^i\right) \beta^i \ .$$

When i is divisible by $r = p - 1$ we get

$$\sum_{\sigma \in G} \varepsilon(\sigma)^i = p - 1 \ .$$

Since $\varphi = \varphi_A \circ \alpha$, we have

$$\varphi_*\big(\mathrm{tr}(\beta^i)\big) = (p-1)\varphi_{A_*}(\beta^i) \neq 0 \ .$$

Hence

$$\varphi_* : K_{2i}(\mathbb{Z}; \mathbb{Z}/p) \longrightarrow \pi_{2i}(BU; \mathbb{Z}/p) = \mathbb{Z}/p$$

is non zero when $p - 1$ divides i.

Corollary 1.4. *When $i \geq 2$ is even the order of $K_{2i-1}(\mathbb{Z})$ is divisible by* $\mathrm{den}(b_i/i)$.

Proof. Let $i = j(p^\nu - p^{\nu-1})$ with j prime to p and $\nu \geq 1$. Consider the commutative diagram

$$
\begin{array}{ccccccccc}
0 & \longrightarrow & K_{2i}(\mathbb{Z})/p^\nu & \longrightarrow & K_{2i}(\mathbb{Z}; \mathbb{Z}/p^\nu) & \longrightarrow & K_{2i-1}(\mathbb{Z})[p^\nu] & \longrightarrow & 0 \\
 & & \downarrow & & \downarrow & & \downarrow & & \\
0 & \longrightarrow & \pi_{2i}(BU)/p^\nu & \longrightarrow & \pi_{2i}(BU; \mathbb{Z}/p^\nu) & \longrightarrow & \pi_{2i-1}(BU)[p^\nu] & \longrightarrow & 0
\end{array}
$$
$$(1.4.1)$$

We know from Gillet's lecture No.6, 6.3., that the map $K_{2i}(\mathbb{Z}) \longrightarrow \pi_{2i}(BU) = \mathbb{Z}$ is zero. Furthermore $\pi_{2i-1}(BU) = 0$. By Theorem 1.1. the map $K_{2i}(\mathbb{Z}, \mathbb{Z}/p^\nu) \longrightarrow \pi_{2i}(BU; \mathbb{Z}/p^\nu)$ is surjective. By chasing in the diagram (1.4.1) we conclude that $K_{2i-1}(\mathbb{Z})$ contains an element of order p^ν. From von Staudt's Theorem (IV 2.3.) the Corollary follows.

2. Numerators

2.1. The numerators of Bernoulli numbers do occur in geometric topology (see for instance [1]). However, up to now, it has not been possible to use purely topological methods to relate these numbers with the K-theory of $\mathbb{Z}$. But methods from algebraic geometry, which make use of the étale topology of the scheme Spec $(\mathbb{Z}[1/p])$, provide such a link.

In [21] Grothendieck showed that the étale cohomology of the classifying (simplicial) scheme BGL_N leads to new Chern classes for representations of discrete groups. These are finer than the usual Chern classes (considered in Chapter IV). Their existence implies for instance that, with the notations of IV.2., $2 \operatorname{den}(b_i/i)\omega^*(c_i) = 0$ ([21], Theorem 4.8). In [41], I used Grothendieck Chern classes to produce new torsion elements in $K_m(\mathbb{Z})$, whose order is related to numerators of Bernoulli numbers. A more powerful formalism was developed by Dwyer and Friedlander [16], and we shall briefly describe their results.

2.2. Let A be a noetherian unitary commutative ring, and p a prime integer invertible in A. Dwyer and Friedlander defined a topological space $BGL(A)^{\text{et}}$, whose homotopy groups are called the *étale K-theory* of A

$$K_m^{\text{et}}(A) = \pi_m\big(BGL(A)^{\text{et}}\big) \ .$$

These are modules over the ring of p-adic integers

$$\mathbb{Z}_p = \varprojlim_n (\mathbb{Z}/p^n) \ .$$

There is a natural map $\rho : BGL(A)^+ \longrightarrow BGL(A)^{\text{et}}$, hence morphisms

$$\rho_m : K_m(A) \underset{\mathbb{Z}}{\otimes} \mathbb{Z}_p \longrightarrow K_m^{\text{et}}(A) \ .$$

When $A = \mathbb{C}$, $BGL(\mathbb{C})^{\text{et}}$ is the p-adic completion of $BGL(\mathbb{C})$ and the map

$$\rho : BGL(\mathbb{C})^+ \longrightarrow BGL(\mathbb{C})^{\text{et}} = BU_p^\wedge$$

is the one induced by $\psi : BGL(\mathbb{C})^+ \longrightarrow BGL(\mathbb{C})$ (III 1.2.).

The groups $K_m^{\text{et}}(A)$ can be computed by means of a spectral sequence $E_r^{st}(A)$ converging to $K_{-s-t}^{\text{et}}(A)$ (when A has finite étale cohomological dimension at p) and such that $E_2^{st}(A) = H_{\text{et}}^s(A; \mathbb{Z}_p(i))$ when $t = -2i \leq 0$, and $E_2^{st}(A) = 0$ otherwise. Here $H_{\text{et}}^s(A, \mathbb{Z}_p(i))$ denotes the étale cohomology of the scheme Spec(A) with coefficients in the i-th Tate twist of $\mathbb{Z}_p$ (a definition when $A = \mathbb{Z}[1/p]$ will be given in 2.4.).

Theorem 2.3. [16]. *For every prime p and every integer $m > 0$ the cokernel of the map*

$$K_m(\mathbb{Z}) \underset{\mathbb{Z}}{\otimes} \mathbb{Z}_p \longrightarrow K_m\big(\mathbb{Z}[1/p]\big)^{\text{et}}$$

is killed by 2.

It is known that $H^k(\mathbb{Z}[1/p], \mathbb{Z}_p(i)) = 0$ when $i \neq 0$ and $k \neq 1, 2$. Hence the spectral sequence $E_r^{st}(\mathbb{Z}[1/p])$ degenerates and

$$K_{2i-1}^{\text{et}}(\mathbb{Z}[1/p]) = H_{\text{et}}^1(\mathbb{Z}[1/p], \mathbb{Z}_p(i)) \qquad i > 0 \ ,$$

and

$$K_{2i-2}^{\text{et}}(\mathbb{Z}[1/p]) = H_{\text{et}}^2(\mathbb{Z}[1/p], \mathbb{Z}_p(i)) \qquad i > 1 \ .$$

When i is even these groups are related to Bernoulli numbers. In fact it is easy to see that, up to a factor 2, $\#H_{\text{et}}^1(\mathbb{Z}[1/p], \mathbb{Z}_p(i))$ is the p-th component of $\text{den}(b_i/i)$ (hence Theorem 2.3 recovers Theorem 1.1.). Furthermore, the proof of the main conjecture on cyclotomic fields [28], together with its translation into étale cohomology [15] show the following:

Theorem 2.4. *Assume $p \neq 2$ and let $i \geq 2$ be an even integer. Then the exact power of p in b_i/i is*

$$(b_i/i)_{p-\text{part}} = \frac{\#H^2(\mathbb{Z}[1/p], \mathbb{Z}_p(i))}{\#H^1(\mathbb{Z}[1/p], \mathbb{Z}_p(i))} \ .$$

If we combine Theorems 2.3. and 2.4. we get that the numerator of b_i/i divides the order of $K_{2i-2}(\mathbb{Z})$ (notice that this numerator is odd).

2.4. We shall not prove Theorem 2.3. and 2.4. but let us at least define $H_{\text{et}}^k(\mathbb{Z}[1/p], \mathbb{Z}_p(i))$. Let $F/\mathbb{Q}$ be a finite extension, and $\mathcal{O}_F$ the ring of integers in F. We say that $F/\mathbb{Q}$ is unramified outside p when, given any prime integer $\ell \neq p$, the ideal ℓ in $\mathcal{O}_F$ is a product

$$(\ell) = \wp_1 \ldots \wp_n$$

of distinct prime ideals. For instance, for every $\nu \geq 1$, the cyclotomic extension of $\mathbb{Q}(\mu_{p^\nu})$ is unramified outside p ([11] 5.2.2., Theorem 2).

Choose an algebraic closure $\bar{\mathbb{Q}}$ of $\mathbb{Q}$ and denote by $\Phi \subset \bar{\mathbb{Q}}$ the union of all finite extensions $F/\mathbb{Q}$, with $F \subset \bar{\mathbb{Q}}$, which are unramified outside p. The maximal p-cyclotomic extension

$$\mathbb{Q}_\infty = \bigcup_{\nu \geq 1} \mathbb{Q}(\mu_{p^\nu})$$

is a subfield of Φ. The cyclotomic character gives an isomorphism $\text{Gal}(\mathbb{Q}_\infty/\mathbb{Q}) \simeq \mathbb{Z}_p^*$, and we denote by

$$\varepsilon : \text{Gal}(\Phi/\mathbb{Q}) \longrightarrow \mathbb{Z}_p^*$$

its composite with the restriction map

$$\text{Gal}(\Phi/\mathbb{Q}) \longrightarrow \text{Gal}(\mathbb{Q}_\infty/\mathbb{Q}) \ .$$

The $\text{Gal}(\Phi/\mathbb{Q})$-module $\mathbb{Z}_p(i)$ is defined as follows. It is equal to $\mathbb{Z}_p$ as abelian group and the action of $g \in \text{Gal}(\Phi/\mathbb{Q})$ is the multiplication by $\varepsilon(g)^i \in \mathbb{Z}_p^*$. For every $k \geq 0$, the group $H_{et}^k \left(\mathbb{Z}[1/p], \mathbb{Z}_p(i) \right)$ is the group of continuous cohomology $H^k \left(\Phi, \mathbb{Z}_p(i) \right)$.

2.5. The study of the relation between algebraic K-theory and étale cohomology (of schemes) has been a very active topic in recent years. Theorem 2.3. is valid for every ring of algebraic integers instead of $\mathbb{Z}$ [16]. Suslin's Theorem 1.1. in Chapter III says that $K_{2i}(\mathbb{C}, \mathbb{Z}/n) = H_{et}^0 \left(\mathbb{C}, \mu_n^{\otimes i} \right)$. For a very general result for schemes see [50].

In low degrees, striking results are known for fields. Let F be an arbitrary field and $n > 1$ an integer invertible in F. Denote by $H^k \left(F, \mu_n^{\otimes i} \right)$ the Galois cohomology groups of F with coefficients in $\mu_n^{\otimes i}$. Let $K_3^{ind}(F)$ be the quotient of $K_3(F)$ by the subgroup generated by products of three elements in $K_1(F) = F^*$.

Theorem 2.5.

  i) [29] $K_2(F)/n$ is isomorphic to $H^2 \left(F, \mu_n^{\otimes 2} \right)$.
  ii) [30] [25] $K_3^{ind}(F)[n]$ is isomorphic to $H^0 \left(F, \mu_n^{\otimes 2} \right)$.
  iii) [30] [25] There is an exact sequence

$$0 \longrightarrow K_3^{ind}(F)/n \longrightarrow H^1 \left(F, \mu_n^{\otimes 2} \right) \longrightarrow K_2(F)[n] \longrightarrow 0 .$$

Assume F is a number field with r_1 real places. By [4] there is an exact sequence

$$(\mathbb{Z}/2)^{r_1} \longrightarrow K_3(F) \longrightarrow K_3(F)^{ind} \longrightarrow 0 .$$

Furthermore $K_3(\mathcal{O}_F) = K_3(F)$ [43], where $\mathcal{O}_F$ denotes the ring of integers in F. When combined with Theorem 2.5. ii) these facts lead to the computation of $K_3(\mathcal{O}_F)$. For instance ([30], 4.19.1.) one recovers $K_3(\mathbb{Z}) = \mathbb{Z}/48$ and one gets the new result $K_3 \left(\mathbb{Z}[\sqrt{-1}] \right) = \mathbb{Z} \oplus \mathbb{Z}/24$.

3. Conclusions

From our (brief) discussion of the current status of the relation between $K_m(\mathbb{Z})$ and the value of the zeta function we saw that a way to go from one to the other consists in two steps:

Step 1. Relate K-theory to étale cohomology. This is an algebraic problem, which is not specific to $\mathbb{Z}$.

Step 2: Relate étale cohomology of $\mathbb{Z}[1/p]$ to values of the zeta function. This is an arithmetic question (relating algebraic and analytic number theory), and the proof of the main conjecture on cyclotomic fields [28] has settled it.

But one might still hope for a more direct link between these two objects, which would come from the use of analytic methods in the study of algebraic K-groups of arithmetic varieties. It was shown by Minkowski that the volume of a fundamental domain for the action of $SL_N(\mathbb{Z})$ on the space of positive definite real quadratic forms (cf. Chapter II) is a product of values of the zeta function at positive integers [33]. Using the modern approach to this fact (Tamagawa numbers), Borel could express the leading term of $\zeta(s)$ at any even negative integer as a "regulator" involving the torsion free part of a K-group of $\mathbb{Z}$ (up to a non zero rational number) [9]. Whether "analytical methods" could also be applied to compute the torsion in the K-theory of $\mathbb{Z}$ is one of the many open problems in the field.

References

[1] Adams, J.F.: On the groups $J(X) - III$. Topology **3** (1965) 193–222
[2] Arlettaz, D.: Chern-Klassen von ganzzahligen und rationalen Darstellungen diskreter Gruppen. Math. Zeits. bf 187 (1984) 49–60
[3] Ash, A.: Small-dimensional classifying spaces for arithmetic subgroups of general linear groups. Duke Math. J. **51**, No. 2 (1984) 459–468
[4] Bass, H., Tate, J.: The Milnor ring of a global field. In: Algebraic K-Theory II. Lecture Notes in Maths. No. 342 (1973) 349–446. Springer-Verlag
[5] Beilinson, A.: Higher regulators and values of L-functions. Journal Soviet Mathematics **30** (1985) 2036–2070
[6] Bloch, S.: Lectures on algebraic cycles. Duke University Mathematics series 4 (1980)
[7] Borel, A.: Introduction aux groupes arithmétiques (1965) Hermann ed.
[8] Borel, A.: Stable real cohomology of arithmetic groups. Ann. Scient. Ec. Norm. Sup. 4ème série, t. 7 (1974) 235–272
[9] Borel, A.: Cohomologie de SL_n et valeurs de fonctions zêta. Ann. Sci. Scuola Norm. Sup. Pisa 4 (1977) 613–636
[10] Borel, A., Serre, J.P.: Corners and arithmetic groups. Comm. Math. Helv. 48 (1974) 244–297
[11] Borevich, I.I., Shafarevich, I.R.: Number Theory. Academic Press, 1966
[12] Bott, R.: The stable homotopy of the classical groups. Annals of Maths. Vol. 70, No. 2 (1959) 313–337
[13] Browder, W.: Algebraic K-theory with coefficients $\mathbb{Z}/p$. Lecture Notes in Math. No. 657 (1978), 40–85. Springer-Verlag
[14] Cartan, H., Eilenberg, S.: Homological Algebra. Princeton University Press, 1956
[15] Coates, J., Lichtenbaum, S.: On ℓ-adic zeta functions. Annals of Maths. **98**, No. 3 (1973) 498–550
[16] Dwyer, W., Friedlander, E.: Algebraic and etale K-theory, Trans. Amer. Math. Soc. **272** (1985) 247-280
[17] Eckmann, B., Mislin, G.: Chern classes of group representations over a number field. Compositio Math. **44** (1981) 41–65
[18] Evens, L., Kahn, D.S.: Chern classes of certain representations of symmetric groups. Trans. AMS **245** (1978) 309–330
[19] Gabber, O.: K-Theory of henselian pairs, preprint (1984)
[20] Gillet, H., Thomason, R.: The K-theory of strict Hensel local rings and a theorem of Suslin. Journal of pure and appl. Algebra **34** (1984) 241–254
[21] Grothendieck, A.: Classes de Chern des représentations de groupes discrets. In: Dix exposés sur la cohomologie des schémas. North-Holland, Masson, 1968
[22] Iwasawa, K.: Lectures on p-adic L-functions. Annals of Math. Studies **74** (1972), Princeton University Press
[23] Lee, R., Szczarba, R.H.: The group $K_3(\mathbb{Z})$ is cyclic of order 48. Annals of Math. **104** (1976), 31–60

[24] Lee, R., Szczarba, R.H.: On the torsion in $K_4(\mathbb{Z})$ and $K_5(\mathbb{Z})$. Duke Math. Journal **45** No. 1 (1978) 101–130, with an Addendum by Soulé, C., p. 131–132

[25] Levine, M.: The indecomposable K_3 of fields. Ann. Ec. Norm. Sup. **22** (1989) 255–344

[26] Lichtenbaum, S.: Values of zeta functions, étale cohomology, and algebraic K-theory. In: Algebraic K-Theory II. Lecture Notes in Maths. No. 342 (1973) 489–501. Springer-Verlag

[27] Loday, J.L.: K-Théorie algébrique et représentations de groupes. Ann. Sci. Ec. Norm. Sup., Série 4, 9 (1976) 309–377

[28] Mazur, B., Wiles, A.: Class fields of abelian extensions of $\mathbb{Q}$. Invent. Math. **76** (1984) 179–330

[29] Merkurjev, A.S., Suslin, A.A.: K-cohomology of Severi-Brauer varieties and norm residue homomorphism. Izv. AN USSR, 46, No. 5 (1982) 1011–1046

[30] Merkurjev, A.S., Suslin, A.A.: On the K_3 of a field. Preprint LOMI (1987)

[31] Milnor, J., Stasheff, J.: Characteristic classes. Annals of Maths. Studies **76** (1974) Princeton University Press

[32] Milnor, J.: Introduction to algebraic K-theory. Annals of Maths. Studies **72** (1971) Princeton University Press

[33] Minkowski, H.: Diskontinuitätsbereich für arithmetische Äquivalenz. Gesammelte Abhandlungen **2** (1905) 53–100

[34] Quillen, D.: Higher Algebraic K-Theory. International Congress of Mathematicians, Vancouver (1974) 171–176

[35] Quillen, D.: Higher algebraic K-theory I. In: Algebraic K-Theory I. Lecture Notes in Maths. No. 341 (1973) 85–147. Springer-Verlag

[36] Quillen, D.: Finite generation of the groups K_i of algebraic integers. In: Algebraic K-Theory I. Lecture Notes in Maths. No. 341 (1973) 178–198. Springer-Verlag

[37] Quillen, D.: Letter to Milnor, July 26, 1972. In: Lecture Notes in Mathematics No. 551 (1976) 182–188

[38] Serre, J.P.: Arbres, amalgames, SL_2. Astérisque **46** (1977)

[39] Soulé, C.: The cohomology of $SL_3(\mathbb{Z})$. Topology **17** (1978) 1–22

[40] Soulé, C.: Groupes arithmétiques et K-théorie des anneaux d'entiers de corps de nombres. Thèse d'état (1978) Université de Paris VII

[41] Soulé, C.: K-Théorie des anneaux d'entiers de corps de nombres et cohomologie étale. Invent. Math. **55** (1979) 251–295

[42] Soulé, C.: Classes de torsion dans la cohomologie des groupes arithmétiques. C.R. Acad. des Sc. Paris **284** (1977) 1009–1011

[43] Soulé, C.: Groupes de Chow et K-théorie de variétés sur un corps fini. Math. Ann. **268** (1984) 317–345

[44] Soulé, C.: K-Théorie et zéros aux points entiers de fonctions zêtas. International congress of Mathematicians, Warsaw (1983) 437–445

[45] Staffeldt, R.E.: Reduction theory and K_3 of the Gaussian integers. Duke Math. Journal **45**, No. 4 (1979) 773–791

[46] Suslin, A.A.: Stability in algebraic K-theory. Lecture Notes in Math. No. 966 (1982) 344–356

[47] Suslin, A.A.: On the K-theory of algebraically closed fields. Invent. Math. **73** (1983) 241–245

[48] Suslin, A.A.: On the K-theory of local fields. J. of pure and appl. algebra **34** (1984) 301–318

[49] Tate, J.: Relations between K_2 and Galois cohomology. Invent. Math. **36** (1976) 257–274

[50] Thomason, R.W.: Algebraic K-theory and étale cohomology. Ann. Sc. Ec. Norm. Sup. **18**, No. 3 (1985) 437–552

CNRS and IHES.

Applications of Group Cohomology to Bilinear Forms

Victor Snaith

Britton Professor of Mathematics, McMaster University, Hamilton, Ontario, Canada

Introduction

These lectures were delivered at UNAM on the occasion of its 75th anniversary of its reopening as part of a series of lectures on aspects of mathematics whose central theme was algebraic K-theory.

Algebraic K-theory, since the work of Quillen, has been intimately connected with group cohomology and with number theory. In view of these considerations it seemed a natural step to the subject of these lectures. They are concerned with relations between characteristic classes, whose values lie in group cohomology or Galois cohomology, some of which are of arithmetic origins and others of which are topological.

These results, originally due to Serre and Frohlich are given in Ch. V (with applications in Ch. VI). The proofs are new. The earlier chapters cover the abelian and non-abelian cohomological background and the principal examples of Galois descent – bilinear forms and central simple algebras.

It was an honour and a pleasure to be invited to give these lectures. I am very grateful to Emilio Lluis-Puebla for his assistance in the preparation of these notes.

I. Abelian Cohomology of Groups

Let G be a group with the discrete topology and M a G-module. Consider the set of functions

$$f : G^n = G \times \ldots \times G \longrightarrow M$$

which is denoted by $C^n(G, M)$ and call it the *n-cochains on G with values in* M. Define a homomorphism

$$d^n : C^n(G, M) \longrightarrow C^{n+1}(G, M)$$

by

$$\begin{aligned}
d^n(f)(g_1, \ldots, g_{n+1}) = {}& g_1 f(g_2, \ldots, g_{n+1}) \\
& + \sum_{i=1}^{n} (-1)^i f(g_1, \ldots, g_i g_{i+1}, \ldots, g_n) \\
& + (-1)^{n+1} f(g_1, \ldots, g_n) \ ,
\end{aligned}$$

where $g_1, g_2, \ldots, g_{n+1} \in G$.

It is easy to check that $d^{n+1} \cdot d^n = 0$ hence $C^*(G, M) = \{C^n(G, M), d^n\}$ is a cochain complex whose n-th cohomology group $(n \geq 0)$ is given by

$$H^n(G, M) = \ker d^n / \operatorname{im} d^{n-1} \ .$$

Here we set $C^n(G, M) = 0$ for $n < 0$.

We will call $H^n(G, M)$ the *abelian cohomology of G* with coefficients in M. The term abelian comes from the fact that M is a G-module, hence an abelian group.

Now, let G be a topological group and M a topological G-module. Let $C_{ct}^n(G, M)$ denote the set of all functions $f \in C^n(G, M)$ such that f is continuous. It is clear that $d_{ct}^n : C_{ct}^n(G, M) \longrightarrow C_{ct}^{n+1}(G, M)$ preserves continuous cochains, hence $C_{ct}^*(G, M)$ is a cochain complex and its cohomology in degree n is

$$H_{ct}^n(G, M) = \ker d_{ct}^n / \operatorname{im} d_{ct}^{n-1}$$

which is called the *continuous cohomology of G with coefficients in M*.

Suppose K is a field and L/K a finite Galois extension of fields. Let $G(L/K)$ denote the Galois group of L/K. Let L', L, K be fields such that $L' \supset L \supset K$ is a chain of finite Galois extensions. Then we have an epimorphism

$$\varphi_{L'/L} : G(L'/K) \twoheadrightarrow G(L/K) \ .$$

Define Ω_K to be the inverse limit over L/K of the groups $G(L/K)$, i.e.

$$\Omega_K = \varprojlim_{L/K} G(L/K) = \left\{ (x_L \in G(L/K)) \, \big| \, \varphi_{L'/L}(x_{L'}) = x_L \right\}$$

where L/K runs over finite Galois extensions within a fixed algebraic closure of K.

Ω_K is contained in the $\Pi_{L/K} G(L/K)$ which is a compact group with the product topology of the discrete topologies on the $(G(L/K))$, hence Ω_K has the resulting subspace topology. Ω_K is compact and totally disconnected. We will call Ω_K *the absolute Galois group of a field* K.

Example. If M has the discrete topology and $f \in C_{ct}^1(\Omega_K, M)$ then f is continuous if and only if there is a factorization

$$
\begin{array}{ccc}
\Omega_K & \xrightarrow{\;f\;} & M \\[4pt]
=\downarrow & & \uparrow \\[4pt]
\Omega_K & \twoheadrightarrow & G(L/K)
\end{array}
$$

We define the *n-th Galois cohomology group of* K *with coefficients in a G-module* M to be the resulting continuous cohomology of the absolute Galois group Ω_K, i.e.

$$H^n(K; M) = H_{ct}^n(\Omega_K; M) .$$

Example. (Basic example of Galois cohomology). Let $M = \mathbb{Z}/n$, the n-th roots of unity in the algebraic closure of K. Then $H^1(K; \mathbb{Z}/n) = \hom_{ct}(\Omega_K, \mathbb{Z}/n)$ is isomorphic to $K^*/(K^*)^n$ where $K^* = K - 0$, and $\frac{1}{n} \in K$. The isomorphism

$$e : K^*/(K^*)^n \longrightarrow H^1(K; \mathbb{Z}/n)$$

is given in the following manner:

$$
\begin{array}{ccc}
K^* & \longrightarrow & \{\Omega_K \longrightarrow \mathbb{Z}/n\} \\[4pt]
a & \longrightarrow & \{g \longmapsto g(b)/(b)\}
\end{array}
$$

where $b^n = a$.

Now, let us define $H^n(G, M)$ by means of homological algebra. Take, for example, $\mathbb{Z}$, as a trvial G-module and construct a sequence of G-module maps

$$0 \longleftarrow \mathbb{Z} \xleftarrow{\;\varepsilon\;} P_0 \xleftarrow{\;d_0\;} P_1 \longleftarrow P_2 \xleftarrow{\;d_2\;} \ldots$$

where each $P_i = \oplus \mathbb{Z}[G]$, that is, a free G-module, such that $\ker d_{i-1} = \operatorname{im} d_i$, ε onto and $\ker \varepsilon = \operatorname{im} d_0$. Apply the $\hom_G(\_, M)$ function to get a cochain complex whose n-th cohomology is

$$H^n(G, M) = \frac{\ker\left(\hom_G(P_n, M) \longrightarrow \hom_G(P_{n+1}, M)\right)}{\operatorname{im}\left(\hom_G(P_{n-1}, M) \longrightarrow \hom_G(P_n, M)\right)} .$$

Such a resolution exists and is called the *bar resolution*. For this we will denote P_n by $B_n G$ which is a free left $\mathbb{Z}[G]$-module on G^n such that

$$\hom_G(B_n G, M) = C^n(G, M) .$$

Let us define some products in cohomology. If M and M' are G-modules, then there is an *internal product*

$$H^n(G, M) \otimes H^m(G, M') \longrightarrow H^{n+m}(G, M \otimes M')$$

given by

$$(f, h) \longmapsto f \otimes h$$

where $f : B_n G \longrightarrow M$, $h : B_m G \longrightarrow M'$ such that $f d_n = 0$ and $h d_n = 0$. Here

$$f \otimes h : B_{n+m} G \xrightarrow{\Delta_{n+m}} \bigoplus_{n+m=a} (B_n G \otimes B_m G) \longrightarrow B_n G \otimes B_m G \longrightarrow M \otimes M' .$$

Explicitly, if the generators of the free $\mathbb{Z}[G]$-module $B_n G$ are denoted by n-tuples of the form $[g_1, \ldots, g_n]$ $(g_1 \in G)$

$$\Delta_t[g_1, \ldots, g_t] = \sum_{a=0}^{t} [g_1, \ldots, g_a] \otimes (g_1 \ldots g_a)[g_{a+1}, \ldots, g_t] .$$

For example, let $M = M' = R$ be a ring with G acting trivially. We have a product $\mu : R \times R \longrightarrow R$ so that we may form

$$
\begin{array}{ccc}
H^n(G, R) \otimes H^m(G, R) & \xrightarrow{(-\cdot-)} & H^{n+m}(G, R \otimes R) \\
= \downarrow & & \downarrow \mu_* \\
H^n(G, R) \otimes H^m(G, R) & \xrightarrow{(-\cup-)} & H^{n+m}(G, R)
\end{array}
$$

This product makes

$$\bigoplus_{n \geq 0} H^n(G, R)$$

into a graded ring, $H^*(G, R)$. In this case $n = 1 = m$, we have $H^1(G, R) \cong \hom(G, R)$ and for $f, h \in \hom(G, R)$, $(f \cup h)[g_1, g_2] = f(g_1)h(g_2) \in R$.

Let $\varphi : G \longrightarrow G'$ be a group homomorphism. Then there is a homomorphism res $: H^n(G', M) \longrightarrow H^n(G, M)$ called the *restriction* and is given by

$$f \longmapsto f(\varphi \times \ldots \times \varphi) .$$

The *transfer* or *corestriction* is defined for a subgroup H of G of index m as a homomorphism

$$\text{tr} : H^n(H, M) \longrightarrow H^n(G, M)$$

in the following manner.
If $0 \longleftarrow \mathbb{Z} \xleftarrow{\varepsilon} Q_0 \xleftarrow{d_0} Q_1 \xleftarrow{d_1} \ldots$ is a free $\mathbb{Z}[H]$-resolution of $\mathbb{Z}$, then

$$0 \longleftarrow \mathbb{Z}[G] \otimes_{\mathbb{Z}[H]} \mathbb{Z} \xleftarrow{1 \otimes \varepsilon} \mathbb{Z}[G] \otimes_{\mathbb{Z}[H]} Q_0 \xleftarrow{1 \otimes d_0} \ldots$$

is an exact $\mathbb{Z}[G]$-resolution of $\mathbb{Z}[G] \otimes_{\mathbb{Z}[H]} \mathbb{Z}$. Hence we have an isomorphism, in which M is a $\mathbb{Z}[G]$-module

$$H^n(H; M)$$
$$\cong \frac{\ker \left\{ d : \hom\left(\mathbb{Z}[G] \otimes_{\mathbb{Z}[H]} Q_n, M\right) \longrightarrow \hom\left(\mathbb{Z}[G] \otimes_{\mathbb{Z}[H]} Q_{n+1}, M\right) \right\}}{\mathrm{im} \left\{ d : \hom\left(\mathbb{Z}[G] \otimes_{\mathbb{Z}[H]} Q_{n-1}, M\right) \longrightarrow \hom\left(\mathbb{Z}[G] \otimes_{\mathbb{Z}[H]} Q_n, M\right) \right\}} \quad (*)$$

However there is a $\mathbb{Z}[G]$-module map

$$\tau : \mathbb{Z} \longrightarrow \mathbb{Z}[G] \otimes_{\mathbb{Z}[H]} \mathbb{Z}$$

given by

$$\tau(t) = \sum_{i=1}^{m} x_i \otimes t$$

where the $\{x_i\}$ are G/H coset representatives. We can extend this map to make the following diagram commutative.

$$
\begin{array}{ccccccccc}
0 & \longleftarrow & \mathbb{Z} & \longleftarrow & Q_0 & \longleftarrow & Q_1 & \longleftarrow & \cdots \\
& & \downarrow{\scriptstyle\tau} & & \downarrow{\scriptstyle\Psi_0} & & \downarrow{\scriptstyle\Psi_1} & & \\
0 & \longleftarrow & \mathbb{Z}[G] \otimes_{\mathbb{Z}[H]} \mathbb{Z} & \longleftarrow & \mathbb{Z}[G] \otimes_{\mathbb{Z}[H]} Q_0 & \longleftarrow & \mathbb{Z}[G] \otimes_{\mathbb{Z}[H]} Q_1 & \longleftarrow & \cdots
\end{array}
$$

where the Ψ_i are G-maps. The induced map τ^* composes with the above isomorphism $(*)$ to yield the *transfer homomorphism*

$$\mathrm{tr} : H^n(H, M) \longrightarrow H^n(G, M) \ .$$

For example, if $K \subset L \subset N$ are fields with L/K a separable extension and $l/n \in K$, we have restriction homomorphism in cohomology and a map of fields. These are related by the following commutative diagram.

$$
\begin{array}{ccc}
H^1(K, \mathbb{Z}/n) & \xrightarrow{\ \mathrm{res}\ } & H^1(L, \mathbb{Z}/n) \\
\| & & \| \\
K^*/(K^*)^n & \xrightarrow{\ \mathrm{inc}\ } & L^*/(L^*)^n
\end{array}
$$

where inc is induced by the inclusion of K into L. We also have the transfer or norm map

$$\mathrm{tr} : H^1(L, \mathbb{Z}/n) \longrightarrow H^1(K, \mathbb{Z}/n) \ .$$

Proposition . *(i) The composition*

$$H^n(G,M) \xrightarrow{\text{res}} H^n(H,M) \xrightarrow{\text{tr}} H^n(G,M)$$

is equal to multiplication by the index m of H in G, (i.e. $\text{tr} \cdot \text{res} = m = [G:H]$).
(ii) Let H, K be subgroups of finite index in G and M a $\mathbb{Z}[G]$-module. The composition

$$H^n(H,M) \xrightarrow{\text{tr}} H^n(G,M) \xrightarrow{\text{res}} H^n(K,M)$$

is given by the Double Coset Formula

$$\text{res} \cdot \text{tr} = \sum_{g \in K \backslash G / H} \Psi_g$$

where Ψ_g is the following composite

$$H^n(H,M) \xrightarrow{(g^{-1} - g)^*} H^n(gHg^{-1}, M) \xrightarrow{\text{res}}$$
$$H^n \left(K \cap (gHg^{-1}), M \right) \xrightarrow{\text{tr}} H^n(K,M) \ .$$

(iii) Let M a trivial G-module. The following diagram commutes

$$
\begin{array}{ccc}
H^1(H,M) & \xrightarrow{\text{tr}} & H^1(G,M) \\
\downarrow{\cong} & & \downarrow{\cong} \\
\hom(H,M) & \xrightarrow{I} & \hom(G,M)
\end{array}
$$

where

$$I(f)(g) = \sum_{i=1}^{m} f\big(h(i,g)\big) \ .$$

Here $m = [G:H]$ and, if $g \in G$, then

$$x_1 g = h(i,g) x_{\pi(i)}$$

($\pi \in \sum_m$ is a permutation and $h(i,g) \in H$ for $1 \le i \le m$).

II.

Let G be a discrete group and let $A \xrightarrow{i} B \xrightarrow{j} C$ be an exact sequence of G-modules. Then there is a long exact sequence of cohomology groups (see (Lluis lecture II.1.6)

$$\ldots \longrightarrow H^n(G,A) \longrightarrow H^n(G,B) \longrightarrow H^n(G,C)$$
$$\xrightarrow{\partial} H^{n+1}(G,A) \longrightarrow H^{n+1}(G,B) \longrightarrow H^{n+1}(G,C) \xrightarrow{\partial} \ldots$$

For example, $\mathbb{Z} \xrightarrow{n} \mathbb{Z} \twoheadrightarrow \mathbb{Z}/n$ is an exact sequence and G acts trivially on $\mathbb{Z}$ and $\mathbb{Z}/n$ then we have a long exact sequence

$$\dots \longrightarrow H^n(G, Z) \xrightarrow{n} H^n(G, Z) \longrightarrow H^n(G, \mathbb{Z}/n) \xrightarrow{\partial} H^{n+1}(G, Z) \longrightarrow \dots .$$

We will now give several examples of cohomology rings.

Example 1. Let $G = \mathbb{Z}/m$, the cyclic group of order n, acting trivially on $\mathbb{Z}$. A free $\mathbb{Z}[G]$-resolution of $\mathbb{Z}$ is

$$0 \longleftarrow \mathbb{Z} \xleftarrow{\varepsilon} \mathbb{Z}[G] \xleftarrow{d_0} \mathbb{Z}[G] \xleftarrow{d_1} \mathbb{Z}[G] \xleftarrow{d_2} \mathbb{Z}[G] \xleftarrow{d_3} \dots$$

where $\varepsilon(g) = 1$ for $g \in G$ and $d_{2j}(\ell_{2j}) = (x-1)\ell_{2j-1}$, $d_{2j+1}(\ell_{2j+1}) = \left(\sum_1^n x^i\right)\ell_{2j}$ where ℓ_i generates $\mathbb{Z}[G]$ and $\langle x \rangle = \mathbb{Z}/n$. Hence $H^m(\mathbb{Z}/n; \mathbb{Z})$ is the cohomology of the complex

$$\mathbb{Z} \xrightarrow{0} \mathbb{Z} \xrightarrow{n} \mathbb{Z} \xrightarrow{0} \dots$$

so that

$$H^i(\mathbb{Z}/n, \mathbb{Z}) \cong \begin{cases} \mathbb{Z} & \text{if } m = 0 \\ 0 & \text{if } m > 0 \text{ is odd} \\ \mathbb{Z}/n & \text{if } m \geq 2 \text{ is even} \end{cases}.$$

For example, if $n = p^\alpha$, p a prime number then

$$H^m(\mathbb{Z}/p^\alpha, \mathbb{Z}/p) = \mathbb{Z}/p \quad \text{for all} \quad m \geq 0$$

by a similar argument. This is the additive structure. To obtain the ring structure we let $P_n = \mathbb{Z}[\mathbb{Z}/p^\alpha]\langle \ell_n \rangle$. Suppose we have cocycles

$$f \in \hom_{\mathbb{Z}/p^\alpha}(B_i \mathbb{Z}/p^\alpha, \mathbb{Z}/p)$$

and

$$h \in \hom_{\mathbb{Z}/p^\alpha}(B_j \mathbb{Z}/p^\alpha, \mathbb{Z}/p)$$

in dimensions i and j respectively. Then the product $[f][h]$ is represented in $\hom_{\mathbb{Z}/p^\alpha}(P_n, \mathbb{Z}/p)$ $(n = 1 + j)$ by

$$P_n \xrightarrow{\phi_n} B_n \mathbb{Z}/p^\alpha \xrightarrow{\Delta_n} \overset{n}{\underset{a=0}{\bigoplus}} B_a \mathbb{Z}/p^\alpha \otimes B_{n-a} \mathbb{Z}/p^\alpha \xrightarrow{f \otimes h} \mathbb{Z}/p .$$

Explicitly, one may define $\varphi_n : P_n \longrightarrow B_n(\mathbb{Z}(p^\alpha))$ as

$$\varphi_n(\ell_n) = \begin{cases} \sum_{i_1 \dots i_s} [x^{i_1}, x, x^{i_2}, x, \dots, x^{i_s}, x] & \text{if } n = 2s > 0 \\ \sum_{i_s \dots i_s} [x, x^{i_1}, x, \dots, x^{i_s}, x] & \text{if } n = 2s + 1 \\ [\,\,] & \text{if } n = 0 \end{cases}.$$

From this formula one easily obtains the following result:

Theorem .

$$H^*(\mathbb{Z}/p^\alpha, \mathbb{Z}/p) = \begin{cases} E(v) \otimes P[f] & \text{if } p \neq 2 \text{ or } p = 2, \ \alpha \geq 2 \\ p[v] & \text{if } p = 2, \ \alpha = 1 \end{cases}$$

where $\deg(v) = 1$, $\deg(f) = 2$. $E(v)$ *and* $P(f)$ *are respectively the* $\mathbb{Z}/p$-*exterior and polynomial algebra on one generator.*

Example 2. Let D_{2n} be the dihedral group of order $2n$, that is

$$D_{2n} = \left\{ x, y | x^n = 1 = y^2, xyx = y \right\} \ .$$

For instance, consider the dihedral group of order eight

$$D_8 = \langle s_1, s_2, t | s_1^2 = s_2^2 = 1 = t^2, s_1 s_2 = s_2 s_1, t s_1 t = s_2 \rangle \ .$$

This group is a subgroup of the orthogenal group of 2×2 real matrices, $0_2(\mathbb{R})$.

Let P_*^1 be the $\mathbb{Z}[\langle s_1 \rangle]$-resolution of the integers, $\mathbb{Z}$, which was described in Example 1,

$$P_*^1 : 0 \longleftarrow \mathbb{Z} \longleftarrow P_0^1 \longleftarrow P_1^1 \longleftarrow P_2^1 \longleftarrow \cdots \ .$$

Similarly, let P_*^2 and P_* be the corresponding resolutions when we replace s_1 by s_2 and t respectively.

Then we obtain an explicit resolution

$$0 \longleftarrow \mathbb{Z} \longleftarrow Q_0 \longleftarrow Q_1 \longleftarrow Q_2 \longleftarrow \cdots$$

of $\mathbb{Z}$ by free D_8-modules where

$$Q_n = \bigoplus_{a+b+c=n} \otimes P_b^1 \otimes P_c^2 \ ,$$

that is, we form the tensor product chain complex

$$\left(P_* \otimes P_*^1 \otimes P_*^2, d \otimes 1 \otimes 1 \pm 1 \otimes d \otimes 1 \pm 1 \otimes 1 \otimes d \right) \ .$$

The n-th group

$$\bigoplus_{a+b+c=n} P_a \otimes P_b^1 \otimes P_c^2$$

is a free $\mathbb{Z}[D_8]$-module if we endow it with the action

$$t \left(\ell_a \otimes \ell_b^1 \otimes \ell_c^2 \right) = (-1)^{bc} \left(t\ell_a \right) \otimes \ell_c^1 \otimes \ell_b^2$$

$$s_1 \left(\ell_a \otimes \ell_b^1 \otimes \ell_c^2 \right) = \ell_a \otimes \left(s_1 \ell_b^1 \right) \otimes \ell_c^2$$

$$s_2 \left(\ell_a \otimes \ell_b^1 \otimes \ell_c^2 \right) = \ell_a \otimes \ell_b^1 \otimes \left(s_2 \ell_c^2 \right) \ .$$

Set

$$\varepsilon = \# \left\{ (a,b) \mid a + 2b = n, n \geq 0 \right\}$$

and

$$\varepsilon'_n = \# \{(x,y) \mid x + 2y = n, y \geq 0, x \geq 1\} \ .$$

We have the following result.

Proposition . $\dim_{\mathbb{F}_2} \big(H^n(D_8, \mathbb{Z}/2)\big) = \varepsilon_n + \varepsilon'_n$.

This is the dimension of the n-dimensional part of the $\mathbb{Z}/2$-algebra $\mathbb{Z}/2 [x_1, x_2, w] / (x_1^2 + x_1 x_2)$ where $\deg x_1 = \deg x_2 = 1$ and $\deg w = 2$.

Stiefel-Whitney classes. Let G be a discrete group and assume the G-module M as trivial action. There is a topological method by which to construct the cohomology of a group G with G-module coefficients M. Choose any contractible space EG on which G acts freely. Form the orbit space $BG = EG/G$ then $H^*(G, M)$ is isomorphic to the singular cohomology of the classifying space BG with coefficients in the G-module M, i.e.

$$H^*(G, M) \cong H^*(BG, M) \ .$$

The isomorphism arises from the fact that any two models for EG define the same right hand side of the isomorphism but one model exists whose singular chain complex $(C_*(EG), d)$ is isomorphic, as a G-chain complex, to the bar resolution of G.

If G is any topological group we may define $H^*(BG, M)$. For example, take the orthogonal group

$$O_n(\bar{K}) = \big\{ X \in GL_n \bar{K} \mid XX^t = I_n \big\}$$

where $\bar{K}$ is a separably closed field with characteristic different from 2.

Let

$$O(\bar{K}) = \bigcup_{n \geq 1} O_n(\bar{K})$$

where the inclusion

$$O_n(\bar{K}) \longrightarrow O_{n+1}(\bar{K})$$

is given by the map

$$X \longmapsto \begin{pmatrix} X & 0 \\ 0 & 1 \end{pmatrix} \ .$$

By corrolaries of Suslin [Su] and Karoubi [Ka] we have that

$$H^*\big(O(\bar{K}); \mathbb{Z}/2\big) \cong \mathbb{Z}/2 [w_1, w_2, \ldots] (\deg w_i = i) \ .$$

Usually, Stiefel-Whitney classes are defined by means of $H^*\big(BO(\mathbb{R}), \mathbb{Z}/2\big) \cong \mathbb{Z}/2 [w_1, w_2, \ldots]$, where $O(\mathbb{R})$ is given the classical topology. But in order to avoid going into details about classifying spaces we will define them as follows: if $\rho : G \longrightarrow O_n(\bar{K}) \longrightarrow O(\bar{K})$ is a homomorphism, we define the i-th *Stiefel-Whitney class* of ρ to be $\rho^*(w_i) = \mathrm{SW}_i[\rho] \in H^i(G, \mathbb{Z}/2)$. If $D_{2n} \overset{\rho}{\longrightarrow} O_2(\mathbb{R})$ then set $w = SW_2[\rho] \in H^2(D_{2n}, \mathbb{Z}/2)$. Take $n = 2^\alpha$. Hence D_4, in particular, is isomorphic to $\mathbb{Z}/2 \times \mathbb{Z}/2$. Let $\alpha \geq 2$, then we have homomorphisms

$$x_i : D_{2^{\alpha+1}} \longrightarrow \mathbb{Z}/2 \quad \text{for} \quad i = 1, 2$$

given by

$$x_i \left(x^\epsilon y^\delta \right) = \begin{cases} \delta(\mathrm{mod}\ 2) & \text{if } i = 1 \\ \varepsilon(\mathrm{mod}\ 2) & \text{if } i = 2 \end{cases} .$$

These give cohomology classes by virtue of the isomorphism $H^1(D_{2^n}, \mathbb{Z}/2) \cong \mathrm{hom}(D_{2^n}, \mathbb{Z}/2)$. We have the following result.

Theorem . *Let $n = 2^\alpha(2s + 1)$ then*

(i) $H^*(D_{2n}, \mathbb{Z}/2) \cong H^*(D_{2^{\alpha+1}}, \mathbb{Z}/2)$
(ii) as an algebra

$$H^*(D_{2^{\alpha+1}}, \mathbb{Z}/2) \cong \begin{cases} \mathbb{Z}/2[x_1, x_2, w] / \left(x_2^2 + x_1 x_2 \right) & \text{if } \alpha \geq 2 \\ \\ \mathbb{Z}/2[x_1, x_2] & \text{if } \alpha = 1 \end{cases} .$$

To show that $x_2^2 + x_1 x_2 = 0$ may be seen by using a trick of Adem. Firstly calculate the dimension of $H^2(D_8, \mathbb{Z}/2)$ and show that x_1, x_2 and w generate the cohomology ring of D_8. Therefore, by dimensions, we get a relation of the form $x_2^2 = ax_1 x_2 + bw + cx_1^2$ where the last two terms are seen to be zero by restricting the subgroups. To prove $ax_1 x_2 \neq 0$ use the outer automorphism $\varphi : D_8 \longrightarrow D_8$ given by $\varphi(x) = x$, $\varphi(y) = xy$. One finds that $\varphi^*(x_1) = x_1$, $\varphi^*(x_2) = x_1 + x_2$ so that applying $\varphi^a st$ to the relation reveals that $a = 1$, at least for D_8. However, by means of the surjection from bigger dihedral groups onto D_8 we see that the same relation holds for all such groups.

Using this example it is possible to prove a very well-known fact concerning products in Galois cohomology: Recall that

$$\begin{aligned} H^*(K; \mathbb{Z}/2) &= H^*_{\mathrm{ct}}(\Omega_K, \mathbb{Z}/2) \\ &= \varprojlim_{L/K} H^*(G(L/K), \mathbb{Z}/2) \end{aligned}$$

where the characteristic of K is different from 2. Let $\ell : K^*/(K^*)^2 \longrightarrow H^1(K, \mathbb{Z}/2)$ be the isomorphism which was introduced above. We have the following result:

Theorem . *For all $a \in K^*$, $\ell(a)\ell(a) = \ell(a)\ell(-1)$.*

Proof. If $\sqrt{a} \in K$ then $\ell(a) = 0$. Remember that $\ell(a)(g) = g(\sqrt{a})/\sqrt{a}$. If $\sqrt{a} \notin K$, $t^4 - a \in K[t]$ is separable and its splitting field is $L = K(\xi, 4), \xi^2 = 1, \mu^4 = a$. We have two cases:

(a) $G(L/K) = D_8$ if $\xi \notin K$, and
(b) $G(L/K) = \mathbb{Z}/4$.

As for case (b): $H^*(\mathbb{Z}/4, \mathbb{Z}/2) = E(v) \otimes P[f]$ hence $\ell(a)^2 = 0$ but $\ell(-1) = 0$ since (-1) has a square root in K in this case.

As for case (a): $x(\xi) = -\xi$, $y(\mu) = \xi\mu$. One easily finds that $x_1 \longleftrightarrow \ell(-1)$ and $x_2 \longleftrightarrow \ell(a)$ which yields the relation

$$\ell(a)^2 = \ell(a)\ell(-1) \ .$$

We have the following examples.

Example . $H^*(D_8, \mathbb{Z}) \cong \mathbb{Z}[\alpha, \beta, \nu, \xi]/(2\alpha, 2\beta, 2\nu, 4\xi, a^2 - \alpha\beta, \nu^2 + \beta\xi)$ where $\deg \alpha = 2 = \deg \beta$, $\deg \nu = 3$ and $\deg \xi = 4$.

By calculations of Janet Aisbett, Emilio Lluis-Puebla and myself we have

Example .

$$H^i\left(GLF_{p^\alpha}[t]/(t^2), \mathbb{Z}\right) = \begin{cases} Z & \text{if } i = 0 \\ 0 & \text{if } i = 1 \\ \left(F_{p^\alpha}[t]/(t^2)\right)^* & \text{if } i = 2 \\ 0 & \text{if } i = 3 \\ \left(1 + t\mathbb{F}_{p^\alpha}[t]\right)^* / (t^4)/(1 + \alpha t^2), \alpha \in \mathbb{F}_{p^\alpha} & \text{if } i = 4 \end{cases}$$

III. Non-abelian Cohomology of Groups

Let G be a topological group, X a topological group with G-action, i.e. X is a (not necessarily abelian) topological group. We define *the 0-th cohomology of G with coefficients in X* as the following subgroup of X:

$$H^0(G, X) = X^G = \{x \in X \mid gx = x \quad \text{for all} \quad g \in G\} \ .$$

Also, we define the *1st cohomology of G with coefficients in X* to be the pointed set with base-point corresponding to $f(g) = x^{-1}g(x)$, $x \in X$

$$H^1(G, X) = Z^1(G, X)/\sim$$

where $Z^1(G, X)$ is the set of continuous 1-cocycles of G with coefficients in X contained in $C_c^1(G, X)$ and $\sim$ is the equivalence relation defined as follows: we will say two 1-cocycles $f_1, f_2 \in Z^1(G, X)$ are equivalent, $f \sim f'$, if there exists an element $x \in X$ such that $f(g) = x^{-1}f'(g)g(x)$.

Let $G \propto X$ be the semidirect product of G and X; that is as a set, $G \propto X$ is the product $G \times X$ with multiplication given in the following way: $(g, x)(g_1, x_1) = (gg_1, xg(x_1))$. Observe that $(g, 1)(1, x)(g^{-1}, 1) = (g, g(x))(g^{-1}, 1) = (gg^{-1}, g(x)) = (1, g(x))$, so X is a normal subgroup of $G \propto X$ and we have a short exact sequence $X \longmapsto G \propto X \twoheadrightarrow G$. Now, given $f : G \longrightarrow X$ a 1-cocycle such that $f(gg') = f(g)g(f(g'))$ then it defines a continuous homomorphism $\varphi : G \longrightarrow G \propto X$ given by $\varphi(g) = (g, f(g))$ since $\varphi(gg') = (gg', f(gg')) = (gg', f(g)g(f(g')) = (g, f(g))(g', f(g')) = \varphi(g)\varphi(g')$. Conversely, given a continuous homomorphism $\varphi : G \longrightarrow G \propto X$ of the form $\varphi(g) = (g, f(g))$ then $f : G \longrightarrow X$ is a continuous map satisfying $f(gg') = f(g)g(f(g'))$; $g, g' \in G$. Hence there is a bijection between $Z^1(G, X)$ and the continuous homomorphisms $\varphi : G \longrightarrow G \propto X$ of the form $\varphi(g) = (g, f(g))$ i.e. the bijection associates φ to f. Observe that $(1, x^{-1}) = (1, x)^{-1}$ in $G \propto X$ and that, in $G \propto X$,

$$(1, x^{-1})\,(g, f_2(g))(1, x) = (g, x^{-1}f_2(g))\,(1, x) = (g, x^{-1}f_2(g)g(x))$$

hence we have that $f_1 \sim f_2$ if and only of f_1 and f_2 are conjugate by $(1, x^{-1}) \in G \propto X$.

Corollary . *Given* $z \in H_c^1(G, X)$ *then there is a well-defined homomorphism*

$$(1, z)^* = z^* : H^1(G \propto X, M) \longrightarrow H^1(G, M)$$

where M *is an abelian, topological, trivial* $(G \propto X)$*-module. If* z *is the base-point then* z^* *is the map induced by the natural map* $\lambda : G \longrightarrow G \propto X$ *given by* $\lambda(g) = (g, 1)$.

Proposition . *[Ser]. Let* $X \overset{i}{\longmapsto} Y \overset{j}{\longrightarrow} W$ *be a central extension of groups upon which* G *acts continuously, in which* j *admits a continuous section as a function. Then the following sequence of pointed sets (and groups) is exact:*

$$X^G \longrightarrow Y^G \longrightarrow W^G \overset{\delta}{\longrightarrow} H_c^1(G, X) \longrightarrow H_c^1(G, Y)$$
$$\longrightarrow H_c^1(G, W) \overset{\Delta}{\longrightarrow} H^2(G, X)$$

where

$$\delta(w) = \{g \longrightarrow y^{-1}g(y) \in X\}, y \in Y, j(y) = w ,$$

and

$$\Delta(f : G \longrightarrow W)(g, g^1) = h(g)g(h(g^1)) [h(gg^1)]^{-1} \in X$$

where f *is a 1-cocycle and* h *is a continuous map of sets* $h : W \longrightarrow Y$ *such that* $jh(w) = w$.

We have some applications: Let K be a field with separable closure $\bar{K}$. If L/K is a finite Galois group by $K \subset L \subset \bar{K}$ we denote, as before, its Galois group by $G(L, K)$. If $L \subset F$ is another finite Galois extension we have an extension

$$G(F/L) \longmapsto G(F/K) \twoheadrightarrow G(L/K)$$

and we may form the absolute Galois group

$$\Omega_K = \varprojlim_{K \subset L \subset \bar{K}} G(L/K)$$

(the limit is taken over finite Galois extensions) with the profinite topology Ω_K acts continuously (i.e. in a "locally constant" manner) on $GL_n \bar{K} = \bigcup_{L/K} GL_n(L)$ and also in the subgroups $SL_n \bar{K}$ and $0_n \bar{K}$. Hence we may form the continuous cohomology groups or Galois cohomology groups $H^1_{ct}(\Omega_K; GL_n \bar{K})$, $H^1_{ct}(\Omega_K, 0_n \bar{K})$ and $H^1_{ct}(\Omega_K, SL_n \bar{K})$ which we will abbreviate as $H^1(K, GL_n \bar{K})$, $H^1(K, 0_n \bar{K})$ and $H^1(K, SL_n \bar{K})$.

Proposition . *Let L/K be a finite Galois extension. Then*

$$H^1(G(L/K); GL_n(L)) = \{1\}$$

and consequently, taking limits over L, $H^1(K, GL_n \bar{K}) = \{1\}$.

Corollary . *If L/K is a finite Galois extension then*

$$H^1(G(L/K), SL_n(L)) = \{1\}$$

and

$$H^1(K, SL_n \bar{K}) = \{1\} .$$

Let L/K be a finite Galois extension of fields. Let V be a finite dimensional vector space over the field K and let $\varphi : V \times V \longrightarrow K$ denote fixed, non-degenerate K-bilinear form. Also, denote by V_L the tensor product $V \otimes_\nu L$ and by φ_L the mapping $\varphi \otimes_K L : V_L \times V_L \longrightarrow L$, given by

$$(v_1 \otimes \ell_1, v_2 \otimes \ell_2) \longrightarrow \varphi(v_1, v_2) \ell_1 \ell_2, \ell_i \in L, v_i \in V .$$

Let $0(\varphi_L)$ be the subgroup of $\text{Aut}_L(V_L)$ that preserves φ_L, that is

$$0(\varphi_L) = \{\alpha \in \text{Aut}_L(V_L) \mid \varphi_L(\alpha(x), \alpha(y)) = \varphi_L(x, y) \quad \text{for all} \quad x, y \in V_L\} .$$

Now suppose $\psi : V \times V \longrightarrow K$ is another bilinear form such that $(V_L, \varphi_L) \cong (V_L, \psi_L)$. This means that there exists an $A \in \text{Aut}_L(V_L)$ so that $\varphi_L(u_1, u_2) = \psi_L(A(u_1), A(u_2))$ for $u_1, u_2 \in V_L$. For $g \in G(L/K)$ set $f(g) = A^{-1}g(A) \in \text{Aut}_L(V_L)$, which defines

$$f : G(L/K) \longrightarrow 0(\varphi_L) .$$

This map leads in the orthogonal group because

$$\varphi(A^{-1}g(A)u_1, A^{-1}g(A)u_2) = \psi_L(g(A)u_1, g(A)u_2)$$
$$= \psi_L(g(A(g^{-1}u_1)), g(A(g^{-1}u_2)))$$
$$= \psi_L(g(A(g^{-1}u_1)), g(A(g^{-1}u_2)))$$

(since $G(L/K)$ acts on V_L, $\mathrm{Aut}_L(V_L)$ and $O(\varphi_L)$ by $g(v \otimes a) = v \otimes g(a)$, $(v \in V, a \in L)$ and $g(\alpha)(w) = g\left(\alpha\left(g^{-1}(z)\right)\right), \alpha \in \mathrm{Aut}_L(V_L), z \in V_L.$)

$$= g\left(\psi_L\left(A\left(g^{-1}u_1\right), A\left(g^{-1}u_2\right)\right)\right)$$
$$= g\left(\varphi_L\left(g^{-1}u_1, g^{-1}u_2\right)\right)$$
$$= \varphi_L\left(u_1, u_2\right) \ .$$

Proposition . *The association* $\psi \longrightarrow [f]$ *induces a one-to-one correspondence*

$$\left\{ \begin{array}{l} K\text{-isomorphism classes of non-singular} \\ \text{bilinear forms } \psi \text{ which are} \\ L\text{-isomorphic to } \varphi_L \end{array} \right\} \longleftrightarrow H^1\big(G(L/K), O(\varphi_L)\big) \ .$$

Proof. If we choose B instead of A, $\varphi(u_1, u_2) = \psi_L(Bu_1, Bu_2)$ hence $B^{-1}A \in O(\varphi_L)$. However, $\left(A^{-1}B\right)\left(B^{-1}g(B)\right) g \left(A^{-1}B\right)^{-1} = A^{-1}g(A)$ hence sending ψ to $[f]$ is well-defined.

Conversely, given a 1-cocycle $f : G(L/K) \longrightarrow O(\varphi_L)$ there is an $A \in \mathrm{Aut}_L(V_L)$ such that $f(g) = A^{-1}g(A)$ since $H^1\left(G(L/K), \mathrm{Aut}_L(V_L)\right) = \{1\}$. Hence we may define $\psi_L : V_L \times V_L \longrightarrow L$ by $\psi_L\left(u_1, u_2\right) = \varphi_L\left(A^{-1}u_1, A^{-1}u_2\right)$ and check that $\psi_L(gx, gy) = g\psi_L(x, y)$ so that ψ_L restricts to a bilinear form, ψ on the $G(L/K)$-invariant subset $(V_L \times V_L)^{G(L/K)} = V \times V$. The two above constructions are mutually inverse.

Let Γ be a finite group which acts on the left of the K-vector space V. If $\psi : V \times V \longrightarrow K$ is a fixed Γ-invariant non-singular form such that $\psi(\gamma x, \gamma y) = \psi(x, y)$ (for $\gamma \in \Gamma$) we have the following

Proposition . *Let* $F : \Gamma \times G(L/K) \longrightarrow O(\varphi_L)$ *be given by* $F(\gamma, g) = q(\gamma)f(g)$ *where* $q(\gamma) = A^{-1}\big(\gamma(A(\_))\big) \in O(\varphi_L)$, $\gamma \in \Gamma$. *Then the association* $\psi \longrightarrow [F]$ *establishes a bijection*

$$\left\{ \begin{array}{l} K[\Gamma]\text{-isomorphism classes of} \\ \text{bilinear forms, non-singular} \\ \text{with } \Gamma\text{-action which are} \\ L\text{-isomorphic (not } \Gamma\text{-equivariantly)} \\ \text{to a fixed } \varphi_L \end{array} \right\} \longrightarrow H^1\left(\Gamma \times G(L/K), O(\varphi_L)\right) \ .$$

IV

Let A be a central simple algebra of finite dimension over a field K. For example, $A = M_n(K)$. By Wedderburn's theorem [Ser I] there exist a finite Galois extension L/K such that there is an isomorphism of L-algebras

$$A_L = A \otimes_K L \cong M_n(L)$$

where $\dim_K A = n^2$.

Also, any matrix $Y \in GL_n(L)$ acts on $M_n(L)$ by conjugation, and the L-algebra automorphism depends only on Y modulo factors, $L^* I_n$. This induces an isomorphism

$$PGL_n(L) = GL_n(L)/L^* \cong \text{Alg Aut}_L\big(M(L)\big)$$

$$Y(\text{mod } L^*) \longmapsto \left(Y \_ Y^{-1}\right) \ .$$

We may, then, construct a 1-cocycle $\psi_A : G(L/K) \longrightarrow PGL_n(L)$ by $\psi_A(g) = B^{-1}g(B) \in \text{Alg Aut}_L\big(M_n(L)\big)$. Hence we have the following

Theorem . ψ_A *above establishes the following bijection*

$$\left\{ \begin{array}{l} \textit{Central simple algebras over } K \\ \textit{such that } A_L \cong M_n(L) \textit{ with} \\ \dim_K A = n^2 \end{array} \right\} \leftrightarrow H^1\big(G(L/K); PGL_n(L)\big)$$

$$M_n(L) \cong A_L \longmapsto B^{-1}g(B) = Bg\left(Bg^{-1}(\_)\right) \ .$$

Theorem . *There is a bijection*

$$\left\{ \begin{array}{l} \textit{Central simple algebras } A \textit{ over} \\ K \textit{ with } \Gamma\textit{-action and } A_L \cong M_n(L) \\ \textit{(not } \Gamma\textit{-equivariantly)} \end{array} \right\} \leftrightarrow H^1\big(\Gamma \times G(L/K); PGL_n(L)\big)$$

where Γ acts trivially on $PGL_n(L)$.

We define the *Brauer group of* K, $\text{Br}(K)$, as the set of isomorphism classes of central simple K-algebras subject to the relation that $M_n D$ is equivalent to D. The operation $(\_\otimes_K \_)$ makes this set into a group. This relation is equivalent to that of making $\text{End}_K(K^n) \cong M_n K$ equivalent to K, the unit of the group.

From the central extension

$$L^* \rightarrowtail GL_n(L) \longrightarrow PGL_n(L)$$

we have an exact sequence of sets and groups

$$\ldots \longrightarrow \{1\} \longrightarrow H^1\big(G(L/K); PGL_n(L)\big) \overset{\Delta}{\longrightarrow} H^2\big(G(L/K), L^*\big)$$

in which Δ is onto if $n = [L : K]$. Taking L/K to the limit in $\bar{K}/K$ we obtain an isomorphism

$$\Delta : \mathrm{Br}(K) \xrightarrow{\cong} H^2(K, \bar{K}^a st) .$$

The previous results show us that non-abelian Galois cohomology can be used to classify bilinear forms, central simple K-algebras and central simple $K[\Gamma]$-algebras, in which the general idea is that $H^1(K; \mathrm{Aut}_{\bar{K}}(C))$ classifies "objects over K" which become isomorphic with X when we "extended scalars to $\bar{K}$".

Example . Let K be a field whose characteristic is different from 2. If $b : V \times V \longrightarrow K$ denotes any symmetric, non-degenerate bilinear form of rank n, we may choose a basis $\{\ell_1\}$ for V such that

$$b(\ell_1, \ell_j) = \begin{cases} a_i & i = j \\ 0 & i \neq j \end{cases} .$$

Over the field $L = K\left(\sqrt{a_1}, \sqrt{a_2}, \ldots, \sqrt{a_n}\right)$ we can define the transformation of $V \otimes_K L$ by $A(\ell_i) = \left(1/\sqrt{a_i}\right)\ell_i$ to obtain $b(A\ell_i, A\ell_j) = \delta_{ij}a_i/\left(\sqrt{(A_i a_j)}\right) = \delta_{ij}$. Hence, if we define for $a \in K^*$ a homomorphism

$$\ell(a) : G(\bar{K}/K) = \Omega_K \longrightarrow O_1(1) = \{\pm 1\}$$
$$g \longmapsto \ell(a)(g) = g(\sqrt{a})/\sqrt{a}$$

then the Galois representative for (V, b) is given by the homomorphism $f : \Omega_K \longrightarrow O_n\bar{k}$

$$g \longmapsto \begin{pmatrix} \ell(a_1)(g) & & & \\ & \ell(a_2)(g) & & \\ & & \ddots & \\ & & & \ell(a_n)(g) \end{pmatrix} = A^{-1}g(A) .$$

Since all (V, b) are diagonalizable, one might think that the preceeding homomorphism finishes the subject in a definitive manner. This is not the case as one may discover if one attempts to diagonalize a bilinear form. For example: let L/K denote a finite separable extension of fields and $b : V \times V \longrightarrow L$ denote a non-singular bilinear form over L (not necessarily symmetric) of rank $m = \dim_L V$. We may compose b with the trace of L/K to form a non-degenerate bilinear form over K of rank md where $d = [L : K]$. This is called the *Scharlau transfer of b* and is denoted by

$$\mathrm{Tr}^S_{L/K}(b) : V \times V \longrightarrow L \xrightarrow{\mathrm{trace}} K$$

$$(x, y) \longmapsto \sum_{\substack{i : L \to K \\ i, \text{ a } K-\text{embedding}}} i\big(b(x, y)\big) .$$

Suppose $L = K(\sqrt{c})$, $c \in K$, $\sqrt{c} \notin K$. Let $\langle a \rangle : L \times L \longrightarrow L$ be given by $\langle a \rangle(u, v) = auv$, if $a = x + y\sqrt{c}$ then

$$\mathrm{Tr}^S_{L/K} \langle a \rangle = \begin{pmatrix} 2x & 2yc \\ 2yc & 2xc \end{pmatrix} \sim \begin{pmatrix} 2x & 0 \\ 0 & \frac{2(x^2c - y^2c^2)}{x} \end{pmatrix} .$$

If $x \neq 0$ we get $\langle 2x \rangle + 2\langle (x^2c - y^2c^2)/x \rangle$ and if $x = 0$ we get $\langle yc \rangle + \langle -yc \rangle$. Now we will illustrate a bilinear form which is impossible to diagonalize:

Example . [Fro] Let $G = G(L/K)$ denote the Galois group of a finite Galois extension L/K. Suppose we have a non-singular bilinear form $b : V \times V \longrightarrow K$ and a homomorphism $T : G \longrightarrow O(b)$ where $O(b)$ is the orthogonal group of (V, b). Let $V^* = \hom_K(V, K)$. We will define a non-degenerate bilinear form

$$\beta^*(T, b) : V^* \times V^* \longrightarrow K .$$

Identify V^* with $\hom_K(V, L)^G$. Then, let $\{v_1, \ldots, v_m\}$ be a basis for V and $u_1, \ldots, u_m \in V$ satisfy $b(v_i, u_i) = \delta_{ij}$. For $f, h \in \hom_K(V, L)$ define

$$\beta^*(T, b)(f, h) = \sum_{j=1}^{m} f(v_j)h(u_j) .$$

Now, define $\beta : V \times V \longrightarrow K$ be requiring that

$$\{f \longmapsto \beta^*(\_, f)\} : V^* \longrightarrow (V^*) \cong V$$

and

$$\{x \longmapsto \beta(x, \_)\} : V \longrightarrow V^*$$

be mutually inverse.

Theorem . *The bilinear form* $\beta : V \times V \longrightarrow K$ *is represented in* $H^1(G(L/K); O(b_L))$ *by the homomorphism*

$$T : G(L/K) \longrightarrow O(b) \longrightarrow O(b_L) .$$

Let $N \supset L \supset K$ and N/K, N/L Galois extensions, $\bar{L} = \bar{K}$. Suppose $\rho : G(N/L) \longrightarrow O_n(K)$ is a homomorphism that represents, in $H^1(G(N/K), O_n(\bar{K}))$ a bilinear form $b : V \times V \longrightarrow L$. We can form the induced representation

$$\mathrm{Ind}_{L/K}(\rho) = K[G(N/K)] \otimes_{K[G(N/L)]} K^n$$

which as a homomorphism is

$$\mathrm{Ind}_{L/K}(\rho) : G(N/K) \longrightarrow O_{nd}(K) \longrightarrow O_{nd}(\bar{K}) ,$$

where $d = [L : K]$.

Theorem . $\mathrm{Tr}^S_{L/K}(V, b)$ *is represented by the* 1-*cocycle* $\mathrm{Ind}_{L/K}(\rho)$ *in* $H^1\left(K, O_{nd}(\bar{K})\right)$.

Example . Let ρ be trivial and $n = 1$. $\mathrm{Tr}^S_{L/K}\langle 1 \rangle$ where $\langle 1 \rangle : L \times L \longrightarrow L$, is called the *trace form* of L/K and is represented by $\mathrm{Ind}_{L/K}(1) : G(N/K) \longrightarrow O_d(\bar{k})$,

$$\mathrm{Ind}_{L/K}(\rho) = K[G(N/K)] \otimes_{K[G(N/L)]} K \ .$$

Let $b : V \times V \longrightarrow K$ be a symmetric, non-degenerate bilinear form of rank $n = \dim_K V$ and assume that $\mathrm{char}(K) \neq 2$. Let $C(V, b)$ denote the *Clifford algebra of* (V, b), that is

$$C(V, b) = T(V)/\left(b(x, x)1 = x^2\right), \quad x \in V \ ,$$

where $T(V) = \oplus_{n \geq 0} (V^{\otimes n})$ is the tensor algebra of V. The dimension over K of $C(V, b)$ is 2^n.

Let $V \subset C(V, b)$. Look at the units $x \in C(V, b)^*$. Define $\alpha : C(V, b) \longrightarrow C(V, b)$ by $\alpha(v) = -v$ for $v \in V$ such that $\alpha(x)Vx^{-1} \subset V$. Such set of units is denoted by $\mathrm{Pin}(V, b)$ and is a group under multiplication.

Let $O(V, b)$ be the orthogonal group of (V, b). Define the *reflection perpendicular to* $v \in V$ to be the map

$$T_v(w) = w - \frac{2b(w, v)v}{b(v, v)}, \quad v, w \in V \ .$$

It is known that $O(V, b)$ is generated by the $\{T_v \mid v \in V\}$. Let $\alpha \in O(V, b)$. Then we define the *Spinor norm of* α to be the homomorphism

$$\theta : O(V, b) \longrightarrow K^*/(K^*)^2$$

given by

$$x \longmapsto \theta(x) = \prod_{i=1}^{t} b(v_i, v_i)$$

where $x = T_{v_1} T_{v_2} \ldots T_{v_t}$.

For example, if

$$x = \begin{pmatrix} 0 & 1 \\ 1 & 0 \end{pmatrix} = T_{v_1 - v_2}$$

then $\theta(x) = 2$.

In general, if σ is a permutation on $O(V, b)$ then $\theta(\sigma) = 2^{\mathrm{sgn}(\sigma)}$.

Consider the group $\mathrm{Pin}(V, b)$, then we have a map

$$\pi : \mathrm{Pin}(V, b) \longrightarrow O(V, b)$$

given by $x \longmapsto \left(\alpha(x)\_x^{-1}\right)$ such that $\ker \pi \cong \mathbb{Z}/2\langle \pm 1 \rangle$ which is central. Here $\alpha : C(V) \longrightarrow C(V)$ is the algebra which is equal to minus the identity on V. In particular

$$\mathbb{Z}/2 \rightarrowtail \mathrm{Pin}(V, b) \twoheadrightarrow \mathrm{im}(\pi) = \ker(\theta)$$

is a central extension.

Let $\nu : V \times V \longrightarrow K$, $(\operatorname{char} K \neq 2)$ be a nondegenerate symmetric bilinear form and let L/K be a finite Galois extension. Let $\rho : G(L/K) \longrightarrow O(V, b)$ be an orthogonal Galois representation. Consider the composition

$$G(L/K) \xrightarrow{\rho} O(V, b) \xrightarrow{\theta} K^* / (K^*)^2 \cong H^1(K; \mathbb{Z}/2) \ .$$

Now, since

$$\theta\rho \in \hom\big(G(L/K), H^1(K; \mathbb{Z}/2)\big)$$
$$\cong \hom\big(G(L/K), \mathbb{Z}/2\big) \otimes H^1(K; \mathbb{Z}/2)$$
$$\cong H^1\big(G(L/K), \mathbb{Z}/2\big) \otimes H^1(K; \mathbb{Z}/2)$$

we may define a class $\operatorname{Sp}[\rho] \in H^2(K; \mathbb{Z}/2)$ equal to the image of $\theta\rho$ under $H^1\big(G(L/K); \mathbb{Z}/2\big) \otimes H^1(K; \mathbb{Z}/2) \longrightarrow H^1(K; \mathbb{Z}/2)^{\otimes 2} \xrightarrow{U} H^2(K; \mathbb{Z}/2)$. For example: if

$$\rho : G(L/K) \longrightarrow \mathbb{Z}/2 \langle \begin{pmatrix} 0 & 1 \\ 1 & 0 \end{pmatrix} \rangle \subset O_n(K)$$

then $\operatorname{Sp}[\rho] = \rho \cup \ell(2)$. In general, if we have $\rho : G(N/K) \longrightarrow \sum_n \subset O_n(K)$ then $\operatorname{Sp}[\rho] = (\det \rho) \cup \ell(2)$.

V

We are going to introduce certain characteristic classes of bilinear forms. Suppose $\beta : V \times V \longrightarrow K$ is a symmetric, non-degenerate bilinear form of rank m over a field K whose characteristic is not 2. We can diagonalize (V, β) over K and obtain $(V, \beta) \cong \langle a_1 \rangle \oplus \langle a_2 \rangle \oplus \ldots \oplus \langle a_m \rangle$ where $\langle a_i \rangle : K \otimes K \longrightarrow K$ sends (x, y) to $a_i x y$.

Let $\ell(a_i) \in H^1(K; \mathbb{Z}/2)$. The *$i$-th Hasse-Witt class of* (V, β) is defined by

$$\operatorname{HW}_i(V, \beta) = \sum_{n_1 < n_2 < \ldots < n_i} \ell(a_{n_1}) \ell(a_{n_2}) \ldots \ell(a_{n_i})$$
$$\in H^i(K; \mathbb{Z}/2) \ .$$

If we formally write $\operatorname{HW}(V, \beta) = 1 + \operatorname{HW}_1(\beta) + \operatorname{HW}_2(\beta) + \ldots$ for the total Hasse-Witt class then

$$\operatorname{HW}\big((V_1, \beta_1) \oplus (V_2, \beta_2)\big) = \operatorname{HW}(V_1, \beta_1) \operatorname{HW}(V_2, \beta_2) \ .$$

Let us use group cohomology to construct Hasse-Witt classes. Let K be a field and for each finite Galois extension, L/K, let

$$A_L \xrightarrow{i_L} B_L \xrightarrow{\pi_L} C_L$$

be given central extension such that $G(L/K)$ acts on it and is natural in L. Choose a family of maps (of sets) $s_L : C_L \longrightarrow B_L$ such that $\pi_L \cdot s_L = 1_{C_L}$. Define

$$w_{2,L} : \left(\Omega_K \propto C_L\right)^2 \longrightarrow A_L/(a_L \sim ga_L) := (A_L)_{\Omega_K}$$

by

$$w_{2,L}\big((g,c),(g',c')\big) = s_L(c)g\big(s_L(c')\big)\left[s_L\big(cg(c')\big)\right]^{-1} \in A_L$$

and then take its image in $(A_L)_{\Omega_K}$. We have the following results:

Theorem . *i) For each L/K, $W_{2,L}$ is a 2-cocycle.*
(ii) Up to 1-boundaries, $w_{2,L}$ is independent of $\{s_L\}$.
(iii) $w_{2,L}\big((g,c),(g',c')\big)$ depends only on the class of g,g' in $G(L/K)$.

Suppose Ω_K acts trivially on A_L, then we have a central extension

$$A_L \rightarrowtail \Omega_K \propto B_L \twoheadrightarrow \Omega_K \propto C_L \ .$$

In this case, the 5-term exact sequence yields a homomorphism

$$\delta : \hom\,(A_L, A_L) \cong H^1\,(A_L, A_L) \longrightarrow H^2\,(\Omega_K \propto C_L; A_L)$$

where

$$1 \longmapsto \delta(1) = w_{2,L} \ .$$

Now, let $\beta : V \times V \longrightarrow K$ be a symmetric, non-singular bilinear form over K, (char $K \neq 2$) of rank m. For each Galois extension L/K consider the central extension

$$\mathbb{Z}/2 \rightarrowtail \mathrm{Pin}\,(V_L, \beta_L) \xrightarrow{\pi_L} (\ker\theta_L) \subset O\,(V_L, \beta_L)$$

$(\cup \ker\theta_L = O\,(V_{\bar{R}}, \beta_{\bar{R}}))$. (V, β) is represented by a class in $H^1\,(K; O_n(\bar{K}))$. We have the following

Theorem . *Let $(V, \beta) \in H^1\,(K; O_n(\bar{K}))$ as above and let*

$$(V, \beta)^* : H^2\,(\Omega_K \propto O_n(\bar{K}); \mathbb{Z}/2) \longrightarrow H^2(K; \mathbb{Z}/2) \ .$$

Then $(V, \beta)^* \{w_{2,L}\} = \mathrm{HW}_2(V, \beta)$.

Let $b : V \times V \longrightarrow K$ be a non-singular bilinear form and $T : G(L/K) \longrightarrow O(b) \longrightarrow O(b_R)$ an orthogonal representation. Then there is a bilinear form, $\beta(T, b) : V \times V \longrightarrow K$ in $H^1\,(G(L/K); O(b_L))$. $\beta(T, b)$ is called the *Frohlich's bilinear form*. We have the following theorem due to Frohlich [Fro I]:

Theorem . *(i) $\mathrm{HW}_1\big(\beta(T, b)\big) = \mathrm{HW}_1(b) + \mathrm{SW}_1[T] \in H^1(K; \mathbb{Z}/2)$.*
(ii) $\mathrm{HW}_2\big(\beta(T, b)\big) = \mathrm{HW}_2(b) + \mathrm{SW}_2[T] + \mathrm{HW}_1(b)\mathrm{SW}_1[T]$ in $H^2(K; \mathbb{Z}/2)$.
Here, $\mathrm{SW}_1[T] = \det T : G(L/K) \longrightarrow \{\pm 1\}$.

Corollary Serre [Ser II]). *Let L/K be a finite separable field extension with char$(K) \neq 2$ and let $\beta : V \times V \longrightarrow L$ be a non-singular, symmetric bilinear form which is represented in $H^1\,(K; O_n(\bar{K}))$ by a continuous homomorphism $f : \Omega_L \longrightarrow O_n(K) \subset O_n(\bar{K})$. Then*

(i) $\mathrm{HW}_1\left(\mathrm{Tr}^S_{L/K}(V,\beta)\right) = \mathrm{SW}_1\left(\mathrm{Ind}^{\Omega_K}_{\Omega_L}[f]\right),$ and

(ii) $\mathrm{HW}_2\left(\mathrm{Tr}^S_{L/K}(\beta)\right) = \mathrm{SW}_2\left(\mathrm{Ind}^{\Omega_K}_{\Omega_L}[f]\right) + \mathrm{Tr}_{L/K}(\mathrm{Sp}[f]) + n\ell(2)d_{L/K'}$

where $d_{L/K}$ is the discriminant of L/K. $\mathrm{Tr}^S_{L/K}$ is the Scharlau transfer, $\mathrm{Tr}_{L/K}$ is the cohomology transfer and Ind^G_H is the induced representation.

To sketch the proof of the second part of the Frohlich formula we need some preparation: we need to describe the Galois cohomology representative of $\beta(T,b)$ in $H^1\left(K;O_n(\bar K)\right)$, $n = \mathrm{rank}_K b = \mathrm{rank}_K \beta(T,b)$. Choose a K-basis for V so that $b((x_1,\ldots,x_n),(y_1,\ldots,y_n)) = \sum a_i x_i y_i$. Set

$$a = \begin{pmatrix} \sqrt{a_1} & & & \\ & \sqrt{a_2} & & \\ & & \ddots & \\ & & & \sqrt{a_n} \end{pmatrix} \in GL_n\bar K$$

so that $(\alpha_-\alpha^{-1}) : GL_n\bar K \longrightarrow GL_n\bar K$ sends $O(b_L)\xrightarrow{\simeq}O_n(\bar K)$, and $\alpha : \bar K^n \longrightarrow \bar K^n$ induces a map on Clifford algebras

$$\lambda : C\left(\bar K^n;b_{\bar K}\right) \longrightarrow C\left(\bar K^n, n\langle 1\rangle\right)$$

such that the following diagram commutes:

$$\begin{array}{ccc} \mathrm{Pin}(b_R) & \xrightarrow{\ \lambda\ } & \mathrm{Pin}_n(\bar K) \\ s_b \uparrow\downarrow \pi_b & & s \uparrow\downarrow \pi \\ O(b_{\bar K}) & \xrightarrow{(\alpha_-\alpha^{-1})} & O_n(\bar K) \end{array}$$

where s, s_b are the sections.

How is β classified in $H^1(K,O_n\bar K)$? If $T(g) = A^{-1}g(A)$ then (for $x = (x_1,\ldots,x_n)$, $y = (y_1,\ldots,y_n)$)

$$\beta_{\bar K}(x,y) = b_{\bar K}\left(A^{-1}x, A^{-1}y\right) = \left(\alpha A^{-1}x, \alpha A^{-1}y\right) \ .$$

So β is represented in $H^1\left(K,O_n(\bar K)\right)$ by

$$\begin{aligned} g \longmapsto \alpha A^{-1}g(A\alpha^{-1}) &= \alpha A^{-1}g(A)\alpha^{-1}\alpha g(\alpha^{-1}) \\ &= \left[\alpha T(g)\alpha^{-1}\right]\left[\alpha g(\alpha^{-1})\right] \ . \end{aligned}$$

Write $X(g) = \alpha g(\alpha^{-1})$ which is the 1-cocycle representing b in $H^1\left(K,O_n(\bar K)\right)$ since $b(x,y) = (\alpha x, \alpha y)$. Also write $h(g) = \alpha T(g)\alpha^{-1}$ so that h is a homomorphism $h : \Omega_K \longrightarrow O_n(\bar K)$ with $\mathrm{SW}_i[T] = \mathrm{SW}_i[h] \in H^i(K;\mathbb{Z}/2)$.

Lemma . *With the above notation, $\gamma : \Omega_K \times \Omega_K \longrightarrow \propto O_n(\bar K)$ given by $\gamma(g,g') = (g, h(g)X(g'))$ is a homomorphism. Also, the composition $\gamma \cdot (\mathrm{diag}) : \Omega_K \longrightarrow \Omega_K \propto O_n(\bar K)$ classifies $\beta(T,b)$ in $H^1\left(K;O_n(\bar K)\right)$.*

The rest of the proof follows by evaluating $(\text{diag})^*\gamma^*(\tilde{w}_2) = \text{HW}_2\big(\beta(T,b)\big)$ using the Künneth formula

$$H^2\left(\Omega_K \times \Omega_K; \mathbb{Z}/2\right)$$
$$\cong H^2(K; \mathbb{Z}/2) \oplus \left(H^1(K; \mathbb{Z}/2) \otimes H^1(K; \mathbb{Z}/2)\right) \oplus H^2(K; \mathbb{Z}/2)$$

from which it is seen that

$$\text{HW}_2(\beta) = \text{HW}_2(b) + \text{SW}_2[T] + \text{Sp}[T] + q\text{SW}_1[T]\text{HW}_1(b)$$

where $q = 0$ or 1. Finally it can be shown that $q = 1$. This sketches the proof of Frohlich's formula.

VI

We will now see some applications of the Frohlich and Serre formulae.

Let L/K be a finite separable extension with $\text{char}(K) \neq 2$. Let N be the normal closure of L/K, i.e. if $\bar{K}$ is separable closure of K containing L then

$$N = \langle \varphi(L) \mid \varphi : L \longrightarrow \bar{K}, \quad \text{a homomorphism}\rangle$$

is the field generated by all K-embeddings of L into $\bar{K}$. The extension N/K is Galois and $G(N/K) \subset \sum_n$ acts transitively on the set $\{1, \ldots, n\}$ (by means of the permutation action of $\sum_n$). Let $H = \{h \mid \lambda(h)(1) = (1)\}$ then $L \cong N^H$ and the inclusion homomorphism is given by

$$\lambda = \text{Ind}_H^{G(N/K)}(1) : G(N/K) \longrightarrow \sum_n \subset O_n(\mathbb{Q})$$

the permutation representation given by the (left) action on the set $G(N/K)/H$. We may apply Serre's formula to λ to get:

$$\text{HW}_1\langle L/K \rangle = \text{SW}_1(\lambda) = d_{L/K}$$
$$\text{HW}_2\langle L/K \rangle = \text{SW}_2(\lambda) + \ell(2)d_{L/K} \ .$$

Here, the trace form of L/K is $\text{Tr}_{L/K}^S\langle 1 \rangle = \langle L/K \rangle$.

Application 1: Let K be a number field and let L/K be an extension of order 4 or 5. Consider the diagrams

$$
\begin{array}{ccccc}
G(N/K) & \xrightarrow{\lambda} & \sum_4 & \subset & O_4 \\
=\uparrow & & \uparrow & & \\
G(N/K) & \xrightarrow{\tilde{\lambda}} & \widetilde{\sum_4} & \subset & \text{Pin}_4 \\
\uparrow & & \uparrow & & \\
\mathbb{Z}/2 & & \mathbb{Z}/2 & &
\end{array}
\qquad
\begin{array}{ccccc}
G(N/K) & \xrightarrow{\lambda} & \sum_5 & \subset & O_5 \\
=\uparrow & & \uparrow & & \\
G(N/K) & \xrightarrow{\tilde{\lambda}} & \widetilde{\sum_5} & \subset & \text{Pin}_5 \\
\uparrow & & \uparrow & & \\
\mathbb{Z}/2 & & \mathbb{Z}/2 & &
\end{array}
$$

Then $\tilde{\lambda}$ exists $\Longleftrightarrow$ $SW_2(\lambda) = 0$ $\Longleftrightarrow$ $HW_2\langle L/K\rangle = \ell(2)d_{L/K}$ $\Longleftrightarrow$

$$\langle L/K\rangle = \begin{cases} 2\langle 1\rangle + \langle 2\rangle + \langle 2d_{L/K}\rangle & \text{for } [L:K] = 4 \\ 3\langle 1\rangle + \langle 2\rangle + \langle 2d_{L/K}\rangle & \text{for } [K:K] = 5 \end{cases} .$$

Also

$$HW_1\langle L/K\rangle = SW_1(\lambda) = d_{L/K}$$

and

$$HW_1\left(\langle 2\rangle + \langle 2d_{L/K}\rangle\right) = \ell(2) + \ell(2) + d_{L/K} = d_{L/K}$$

while

$$HW_2\left(\langle 2\rangle + \langle 2d_{L/K}\rangle\right) = \ell(2)\left(\ell(2) + d_{L/K}\right) = \ell(2)d_{L/K} .$$

This means that the trace form $\langle L/K\rangle$ has the same rank and Hasse-Witt classes as its putative diagonalization.

Application 2: Let K be a number field and L/K a triquadratic extension. That is, $L = K\left(\sqrt{x}, \sqrt{y}, \sqrt{z}\right)$ with $xyz = 1$, $[L:K] = 4$ and $G(L/K) \cong \mathbb{Z}/2 \times \mathbb{Z}/2$. Hence the trace form, $\langle L/K\rangle$ is represented by the regular representation, $1 + \ell(x) + \ell(y) + \ell(z)$, that is

$$\langle L/K\rangle = \langle 1\rangle + \langle x\rangle + \langle y\rangle + \langle z\rangle .$$

Let $N = L$, so we also have $\tilde{G}(L/K) = \tilde{G}(N/K) = Q_8$ the quaternion group of order 8. In the following diagram

$$
\begin{array}{ccc}
\mathbb{Z}/2 & \longrightarrow & \mathbb{Z}/2 \\
\downarrow & & \downarrow \\
Q_8 & \xrightarrow{\ \lambda\ } & \tilde{\Sigma_4} \\
\downarrow & & \downarrow \\
\mathbb{Z}/2 \times \mathbb{Z}/2 & \longrightarrow & \Sigma_4
\end{array}
$$

$\tilde{\lambda}$ exists if and only if $\langle x\rangle + \langle y\rangle + \langle z\rangle = 3\langle 1\rangle$. HW_1 and HW_2 characterize forms of rank three over a number field.

Application 3: Suppose $[L:K] = 5$, $d_{L/K} = 1$ and $5 \in (K^*)^2$. Then the following conditions are equivalent.

(a) $SW_2(\lambda) = \ell(-1)^2$.
(b) $\langle L/K\rangle = 3\langle 1\rangle + 2\langle -1\rangle$.
(c) there exists $x \in L^*$ such that $\text{tr}_{L/K}(x) = 0 = \text{tr}_{L/K}(x^2)$.

The conditions (a) and (c) are equivalent to the fact that the extension L/K is constructible by an icosahedral construction due to Klein.

To close this section let me state an old conjecture by Milnor [Mil]. It is known that of K is a number field then

$$K_1(K; \mathbb{Z}/n) \cong K^* \otimes \mathbb{Z}/n \cong H^1(K; \mathbb{Z}/n)$$

where K_1 denotes algebraic K-theory and H^1 denotes Galois cohomology. Also, by a theorem of Matsumoto

$$K_2(K) \cong \frac{K^* \otimes K^*}{\{(a \otimes (1-a))\}} \qquad a \neq 0, 1 .$$

$$(x, y) \longleftarrow x \otimes y .$$

If $\frac{1}{n} \in K$, $\exp\left(\frac{2\pi i}{n}\right) \in K$ then $\ell(a)\ell(1-a) \in nH^2(K; \mathbb{Z}/n)$ and this means that we have a map which is an isomorphism

$$K_2(K) \otimes \mathbb{Z}/n \xrightarrow{\cong} H^2(K; \mathbb{Z}/n)$$

$$x \otimes y \longmapsto \ell(x)\ell(y)$$

proved by Tate for number fields K and for all fields by Merkurjev-Suslin.

Define Milnor's K-theory as

$$K_i^{\mathrm{Mil}}(K) = \frac{\overbrace{K^* \otimes \ldots \otimes K^*}^{n}}{\{\text{relations generated by } a \otimes 1 - a\}} .$$

Then

$$K_i^{\mathrm{Mil}}(K) \otimes \mathbb{Z}/n \longrightarrow H^i(K; \mathbb{Z}/n)$$

$$x_1 \otimes \ldots \otimes x_i \longmapsto \ell(x_1) \ldots \ell(x_i)$$

is an isomorphism for K a number field.

Let $W(K)$ be a Witt ring of symmetric bilinear forms over K and $G(K)$ the Grothendieck group of bilinear forms.

$$W(K) = G(K) / \left\langle \begin{pmatrix} 0 & 1 \\ 1 & 0 \end{pmatrix} \right\rangle .$$

If we modify the "HW"-classes so as to be well-defined on $W(K)$ we get

$$W(K) = G(K) / \langle \begin{smallmatrix} 0 & 1 \\ 1 & 0 \end{smallmatrix} \rangle \begin{array}{cc} \text{"HW}_1\text{"} & H^1(K; \mathbb{Z}/2) \\ \longrightarrow & \\ \text{"HW}_2\text{"} & H^2(K; \mathbb{Z}/2) \end{array}$$

$$\cdots .$$

There is a filtration on $W(K)$, $F^n(W(K))$, such that $F^0/F^1 \cong H^1(K; \mathbb{Z}/2)$, $F^1/F^2 \cong H^2(K; \mathbb{Z}/2)$ and in general, for number fields we have

$$F^i/F^{i+1} \cong H^i(K; \mathbb{Z}/2) \cong K_i^{\mathrm{Mil}}(K) \otimes \mathbb{Z}/2 .$$

Milnor's conjecture is that, for a field K with $\text{char}(K) \neq 2$,

$$F^i/F^{i+1} \longrightarrow K_i^{\text{Mil}}(K) \otimes \mathbb{Z}/2$$

is a isomorphism.

In 1986 Merkurjev-Suslin and, independently, Levine and Rost established the analogous isomorphism in dimension three

$$K_3^{\text{Mil}}(K) \otimes \mathbb{Z}/2 \xrightarrow{\cong} H^3(K; \mathbb{Z}/2)$$

for any field in which 2 is invertible.

VII. Orthogonal Local Root Numbers

The material of this section was not part of my original lectures. However, as the reader will see below, this material uses Galois cohomology and the Grothendieck-Witt ring of non-singular, symmetric bilinear forms over a local field.

In this section, as an application of the orthogonal Explicit Brauer Induction theorem of [Sn 1], I will sketch how to construct orthogonal local root numbers. This is done by proving a formula (Theorem 7.8), which is originally due to Deligne, for the orthogonal local root numbers in terms of the second Stiefel-Whitney class in Galois cohomology.

Firstly let us briefly recapitulate the theory of these local root numbers (sometimes called *local constants*).

Complete details concerning the material of this section are to be found in [Sn 1; Sn 2].

Let K be a non-Archimedean local field. For simplicity we will assume that $\text{char}(K) = 0$. Let Ω_K denote the absolute Galois group of K and suppose that $\nu : \Omega_K \longrightarrow U_n(\mathbb{C})$ is a continuous, unitary Galois representation. To such a ν is associated the Deligne-Langlands root number

$$W_K(\nu) \in S^1 . \tag{7.1}$$

The $\{W_K(\nu)\}$ are particularly important in the theory of Artin L-functions of number fields. For example, the Artin root number of a continuous Galois representation, ρ, of a number field is given as the product over primes, v, of the local root numbers of the local representations, ρ_v, given by restricting ρ to a decomposition group of v.

Suppose that L/K is a finite Galois extension of local fields and that ν is a continuous, orthogonal Galois representation of the form

$$\Omega_K \xrightarrow{\pi} G(L/K) \xrightarrow{\nu} O_n(\mathbb{R}) . \tag{7.2}$$

In (7.2) π is the canonical map onto the finite Galois group, $G(L/K)$, of L/K.

When ν has the form (7.2) then $W_K(\nu)$ is a fourth root of unity. We will write μ_4 for the cyclic group of fourth roots of unity (generated by $i = +(\sqrt{(-1)})$, say). Denote by 2_* the inclusion of the group of order two

$$2_* : \{\pm 1\} \rightarrowtail \mu_4 = \{i\} \ . \tag{7.3}$$

There are isomorphisms

$$\begin{cases} \{\pm 1\} \cong H^2(K; \mathbb{Z}/2) \\ \ell : K^*/(K^*)^2 \xrightarrow{\cong} H^1(K; \mathbb{Z}/2) \end{cases} \tag{7.4}$$

where $H^i(K; \mathbb{Z}/2)$ denotes Galois cohomology. The isomorphism, ℓ, in (7.4), is given as in §I, by

$$\ell(a)(g) = g(\sqrt{a})/(\sqrt{a}), \, (g \in \Omega_K, a \in K^*/(K^*)^2) \ . \tag{7.5}$$

By (7.4) the cup-product becomes

$$\left. \begin{array}{l} K^*/(K^*)^2 \otimes K^*/(K^*)^2 \longrightarrow \{\pm 1\} \\ \\ \ell(a) \otimes \ell(b) \rightarrowtail (\ell(a) \cup \ell(b)) \end{array} \right\} \tag{7.6}$$

Now let F/K be a finite extension of local fields and let $\rho : \Omega_F \longrightarrow O_n(\mathbb{R})$ be a continuous, orthogonal representation. Write

$$\mathrm{Ind}_{F/K}(\rho) : \Omega_K \longrightarrow O_{nd}(\mathbb{R}), \quad (d = [F : K])$$

for the induced representation.

The i-th *Stiefel-Whitney class* (see §II) lies in the i-th Galois cohomology group. For example, if ρ is as above then

$$SW_1(\rho) = \det(\rho) \in H^1(F; \mathbb{Z}/2) \cong F^*/(F^*)^2 \quad \text{and}$$
$$SW_2(\rho) \in H^2(F; \mathbb{Z}/2) \cong \{\pm 1\} \ .$$

(7.7.) The $\{W_K(\nu)\}$ satisfy the following properties:

(i) Let F/K be a finite extension and let $\rho : \Omega_F \longrightarrow O_n(\mathbb{R})$ be a continuous Galois representation then

$$W_K\left(\mathrm{Ind}_{F/K}(\rho)\right) = W_F(\rho) \cdot W_K\left(\mathrm{Ind}_{F/K}(1)\right)^{\dim \rho} \ .$$

(ii) If $a, b \in K^*$ then

$$W_K\left(\ell(ab)\right) = W_K\left(\ell(a)\right) \cdot W_K\left(\ell(b)\right) \cdot \left[2_*\left(\ell(a) \cup \ell(b)\right)\right] \ .$$

(iii) If $\nu : \Omega_K \longrightarrow O_n(\mathbb{R})$ is a continuous Galois representation with $n \leq 2$ then

$$W_K(\nu) = \left(2_*(SW_2[\nu])\right), W_K(\det \nu) \ .$$

We will need the following result, originally proved in a different manner by Deligne.

Set

$$RO(\Omega_K) = \lim_{L/K} RO\big(G(L/K)\big) \ ,$$

the direct limit of the orthogonal representation rings of the finite Galois groups, $G(L/K)$.

Theorem 7.8. *Any family of homomorphisms*

$$W_K : RO(\Omega_K) \longrightarrow \mu_4$$

satisfying §7.7(i)–(iii) must satisfy

$$W_K(\nu) = (2_*(\mathrm{SW}_2[\nu])) \cdot W_K(\det \nu)$$

for any $\nu : \Omega_K \longrightarrow O_n(\mathbb{R})$ as in (7.2).

Proof. Throughout this proof let us write $s(\nu)$ for $2_* (\mathrm{SW}_2[\nu])$.

By §7.7(i), $W_K(1) = 1$ so we may study $W_K(\nu - n)$. Also by §7.7(ii) we may replace ν by $\nu \oplus (\det \nu)$ if necessary. Hence we may suppose that ν is special orthogonal and if the form $\nu : \Omega_K \longrightarrow SO_{2m}(\mathbb{R})$ with $m \geq 2$, by §7.7(iii).

Assume that we have proved the formula for any orthogonal representation of dimension less than or equal to $2m - 2$.

By the orthogonal Explicit Brauer Induction theorem of [Sn 1]

$$\nu - 2m = \sum_\alpha n_\alpha \left(\mathrm{Ind}_{F_\alpha/K} \left((\nu \mid \Omega_{F_\alpha}) - 2m \right) \right)$$

where F_α/K is such that $\nu (\Omega_{F_\alpha})$ lies in the normaliser, NT^m, of the diagonal torus $T^m = SO(2)^m$ in $SO_{2m}(\mathbb{R})$ and $\mathrm{Ind}_{L/K}$ denotes induction of representation from Ω_L to Ω_K.

On special orthogonal representations (or vector bundles) of virtual dimension zero SW_2 is a homomorphism. This homomorphism commutes with induction since SW_2 is the first k-invariant of BSO (see [Sn 1]).

Suppose that

$$W_{F_\alpha} \left((\nu \mid \Omega_{F_\alpha}) - 2m \right) = s \left((\nu \mid \Omega_{F_\alpha}) - 2m \right)$$

then

$$s(\nu - 2m) = \prod_\alpha \left(s \left(\mathrm{Ind}_{F_\alpha/K} \left((\nu \mid \Omega_{F_\alpha}) - 2m \right) \right) \right)^{n_\alpha}$$

$$= \prod_\alpha \left(s \left((\nu \mid \Omega_{F_\alpha}) - 2m \right) \right)^{n_\alpha}$$

$$= \prod_\alpha W_{F_\alpha} \left((\nu \mid \Omega_{F_\alpha}) - 2m \right)^{n_\alpha}$$

$$= \prod_\alpha W_K \left(\mathrm{Ind}_{F_\alpha/K} \left((\nu \mid \Omega_{F_\alpha}) - 2m \right) \right)^{n_\alpha}$$

$$= W_K(\nu - 2m) \ .$$

Therefore we are reduced to studying ρ of the form $\rho : \Omega_K \longrightarrow NT^m \subset SO_{2m}(\mathbb{R})$.

Let $O_2(\mathbb{R}) \wr \sum_m$ denote the wreath product generated by the diagonal $O_2(\mathbb{R})^m$ in $O_{2m}(\mathbb{R})$ and the permutation matrices permute the diagonal 2×2 blocks, $O_2(\mathbb{R})^m$. Suppose that

$$\rho : \Omega_K \longrightarrow NT^m \subset O_2(\mathbb{R}) \wr \sum_m \twoheadrightarrow \sum_m .$$

acts on $\{1, 2, \ldots, m\}$ with orbits of size $m_1, m_2, \ldots, m_r$ then ρ can be written

$$\rho = \sum_{j=1}^{r} \mathrm{Ind}_{N_j/K} \left(\Omega_{N_j} \xrightarrow{\rho_j} O_2(\mathbb{R}) \right)$$

where $[N_j : K] = m_j$. Write ϵ_j for the homomorphism

$$\left. \begin{array}{c} \epsilon_j : \Omega_{N_j} \longrightarrow O_2(\mathbb{R}) \\[2mm] g \mapsto \begin{bmatrix} \det \rho_j(g) & 0 \\ 0 & 1 \end{bmatrix} \end{array} \right\} \tag{7.9}$$

Hence

$$\begin{aligned} W_K(\rho) &= W_K \left(\rho - \sum_j \mathrm{Ind}_{N_j/K}(\epsilon_j) + \sum_j \mathrm{Ind}_{N_j/K}(\epsilon_j) \right) \\[2mm] &= \left(\prod_j W_{N_j} \left(\rho_j - \epsilon_j \right) \right) \left(\prod_j W_K \left(\mathrm{Ind}_{N_j/K} \left(\epsilon_j \right) \right) \right) \\[2mm] &= \left(\prod_j \left(s \left(\rho_j - \epsilon_j \right) \right) \right) \left(\prod_j W_K \left(\mathrm{Ind}_{N_j/K} \left(\epsilon_j \right) \right) \right) \\[2mm] &= \left(\prod_j \left(s \left(\mathrm{Ind}_{N_j/K} \left(\rho_j - \epsilon_j \right) \right) \right) \right) \left(\prod_j W_K \left(\mathrm{Ind}_{N_j/K} \left(\epsilon_j \right) \right) \right) \\[2mm] &= \left(s \left(\rho - \sum_j \mathrm{Ind}_{N_j/K} \left(\epsilon_j \right) \right) \right) \left(\prod_j W_K \left(\mathrm{Ind}_{N_j/K} \left(\epsilon_j \right) \right) \right) \end{aligned}$$

since $\det(\rho_j) = \det(\epsilon_j)$.

Now, as a representation, $e_j = 1 + d_j$ where d_j is given by the determinant of ϵ_j so that

$$\mathrm{Ind}_{N_j/K}(\epsilon_j) = \mathrm{Ind}_{N_j/K}(1) \oplus \mathrm{Ind}_{N_j/K}(d_j) = A_j \oplus B_j, \quad \text{say} .$$

Notice that each of A_j and B_j has dimension less than or equal to $2m - 2$. Hence, by induction,

$$W_K\left(\mathrm{Ind}_{N_j/K}\left(\epsilon_j\right)\right) = s\left(A_j\right)s\left(B_j\right)W_K\left(\det A_j\right)W_K\left(\det B_j\right)$$
$$= s\left(A_j\right)s\left(B_j\right)W_K\left(\det(A_j\oplus B_j)\right)\left[2_*\left(\det A_j\cup\det B_j\right)\right]$$

by §7.7(ii),

$$= s\left(A_j\right)s\left(B_j\right)\left(\det(A_j\oplus B_j)\right)\left[2_*\left(\mathrm{SW}_1(A_j)\right)\cup\left(\mathrm{SW}_1(B_j)\right)\right]$$
$$= s\left(A_j\oplus B_j\right)W_K\left(\det(A_j\oplus B_J)\right)$$

by the Cartan formula.

Similarly, by the iterated application of §7.7(ii) and the Cartan formula, $\mathrm{SW}_2(y+z) = \mathrm{SW}_2(y) + \mathrm{SW}_2(z) + (\mathrm{SW}_1(y)\mathrm{SW}_1(z))$,

$$\prod_j W_K\left(\mathrm{Ind}(\epsilon_j)\right) = s\left(\sum_j\mathrm{Ind}(\epsilon_j)\right)W_K\left(\det\left(\sum_j\mathrm{Ind}(\epsilon_j)\right)\right)$$
$$= s\left(\sum_j\mathrm{Ind}(\epsilon_j)\right)$$

since the determinant above is equal to that of ρ, which is trivial since ρ is special orthogonal. This completes the proof.   □

7.10.

Remark. The properties of §7.7(i)–(iii) are not independent. In fact §7.7(ii) is a consequence of the formula of Fröhlich and Queyrut for dihedral root numbers, which is part of §7.7(iii). I will conclude this section with a sketch of a new construction of orthogonal root numbers, details are to be found in [Sn 2].

7.11. As in §7.7 set

$$RO(\Omega_K) = \varinjlim_{L/K} RO\left(G(L/K)\right)$$

and let $IO(\Omega_K)$ denote the augmentation ideal. There is a filtration on $RO(\Omega_K)$

$$RO(\Omega_K) \supset IO(\Omega_K) = F^1RO(\Omega_K) \supset F^2RO(\Omega_K) \supset F^3RO(\Omega_K)\ldots$$

which is a topological nature. It comes from the filtration associated to the Atiyah-Hirzebruch spectral sequence for the orthogonal K-theory of the classifying spaces of the $\{G(L/K)\}$. The connection between topology and representation theory is afforded by the isomorphism

$$KO(BG(L/K)) \cong RO\left(G(L/K)\right)^{\wedge}$$

where $X^\wedge$ denotes the $IO(G(L/K))$-adic completion of X.

This filtration satisfies $IO(\Omega_K)^3 \subset F^3 RO(\Omega_K)$.

We define

$$Y_K = IO(\Omega_K)/(F^3 RO(\Omega_K)) \ . \tag{7.12}$$

For all $a \in K^*$ we set

$$\hat{p}_K(a) = \ell(a) - 1 \in IO(\Omega_K) \ . \tag{7.13}$$

The following result characterises Y_K.

Theorem 7.14. *Let K be a local field.*

(i) There is an exact sequence

$$\{\pm 1\} \cong H^2(K; \mathbb{Z}/2) \overset{i}{\rightarrowtail} Y_K \overset{\pi}{\twoheadrightarrow} K^*/(K^*)^2 \ .$$

(ii) If $p_K(a) \in Y_K$ is the image of $\hat{p}_K(a)$ $(a \in K^*)$ then

$$p_K(ab) = p_K(a) + p_K(b) + i(\ell(a)\ell(b)) \in Y_K \ .$$

(iii) $\pi(p_K(a)) = \ell(a) \in K^*/(K^*)^2 \cong H^1(K; \mathbb{Z}/2)$.

(iv) Let $\nu : \Omega_K \longrightarrow SO(n)$ be a continuous, special orthogonal Galois representation then the image of $(\nu - n)$ in Y_K is given by $i(\mathrm{SW}_2(\nu)) \in \{\pm 1\}$.

7.15.

Remark. Pierre Conner has pointed out to me that Y_K is isomorphic to the fundamental ideal in the Witt ring of K. If $W(K)$ is the Witt ring of K then the fundamental ideal, $IW(K)$, is the kernel of the map that sends the Witt class of a quadratic form to its rank modulo two. The structure of $IW(K)$ is known and is characterised by an exact sequence, as in Theorem 7.14. In $W(K)$ one replaces $p(a)$ by the classes of the bilinear form

$$K \times K \longrightarrow K \quad \text{given by} \quad ((x, y) \longrightarrow axy) \ .$$

This connection is elaborated upon more seriously in [Sn 1] where I describe the connection between orthogonal local root numbers and the Weil character on the Witt ring of a local field.

7.16. Let K be a non-Archimedean local field containing $\mathbb{Q}_p$. Consider the following commutative diagram

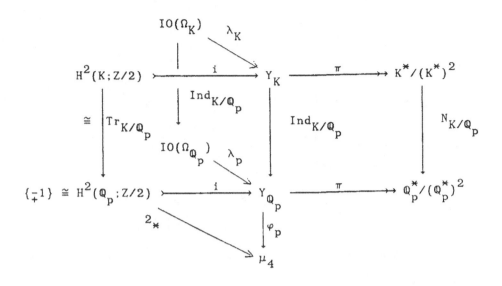

(7.17)

The homomorphism, φ_p, in (7.17) is chosen so that

$$W_{\mathbb{Q}_p}\big(\ell(a)\big) = \varphi_p[\ell(a) - 1] \in \mu_4, \quad (a \in \mathbb{Q}_*) \tag{7.18}$$

where we consider $\ell(a)$ as a one-dimensional, orthogonal Galois representation. An explicit choice for φ_p is tabulated below.

Table A: $\varphi_2[\ell(a) - 1] \in \mu_4$ for $a \in \mathbb{Q}_2^*/\big(\mathbb{Q}_2^*\big)^2 = \{\pm 1; \pm 2, \pm 5, \pm 10\}$

a	1	-1	5	-5	2	-2	10	-10
$\varphi_2\big(\ell(a) - 1\big)$	1	i	1	i	1	i	-1	$-i$

When $p \neq 2$ $\mathbb{Q}_p^*(\big(\mathbb{Q}_p^*\big)^2$ is generated by p and by a unit u whose Legendre symbol $\left[\frac{u}{p}\right]$ is non-trivial.

Table B: $\varphi_p[\ell(a) - 1] \in \mu_4$ for $a \in \mathbb{Q}_p^*/\big(\mathbb{Q}_p^*\big)^2$, $p \neq 2$

a	1	u	p	up
$\varphi_p[\ell(a) - 1]$	1	1	$(p^{-1/2}) \sum_{x \in F_p^*} (-1)^{\epsilon(p)} \left[\frac{x}{p}\right] \cdot \xi_p^x$	$-\varphi_p[\ell(p) - 1]$

In Table B, $\xi_p = \exp(2\pi i/p)$ and $\epsilon(z) \equiv (z - 1)/2 \pmod 2$.

From the Gauss sum/conductor formula for the local root numbers of one-dimensional representations [Sn 1] one easily verifies (7.18). For this purpose one needs §7.14(ii) and the cup products in $H^2(\mathbb{Q}_p; Z/2)$ (the Hilbert symbols) which are to be found in [Ser III, pp. 15–20].

The Gauss sum in Table B is known to satisfy

$$\varphi_p[\ell(p) - 1] = \begin{cases} -i & \text{if } p \equiv 3 \ (\text{mod } 4) \ . \\ +1 & \text{if } p \equiv 1 \ (\text{mod } 4) \ . \end{cases} \tag{7.19}$$

Now we come to the new description of orthogonal root numbers.

Theorem 7.20. *Let $K/\mathbb{Q}_p$ be a finite extension of local fields. Then for*

$$\nu : \Omega_K \xrightarrow{\pi} G(L/K) \longrightarrow O_n(\mathbb{R}) \ ,$$

as in (7.2), defining $W_K(\nu)$ by

$$W_K(\nu) = \varphi_p \left(\lambda_p \left(\operatorname{Ind}_{K/\mathbb{Q}_p}(\nu - n) \right) \right)$$

$$= \varphi_p \left(\operatorname{Ind}_{K/\mathbb{Q}_p} \left(\lambda_K(\nu - n) \right) \right) \in \mu_4 \ ,$$

satisfies the axioms of §7.7.

References

[Fro]   Frohlich, A.: Orthogonal representations of Galois groups, Stiefel-Whitney classes and Hasse-Witt invariants. J.F. Reine. Ang. Math. **360** (1985), 85–123

[Fro I]   Frohlich, A.: Orthogonal and symplectic representations of groups. Proc. L.M. Soc. (3)24 (1972), 470–506

[Ka]   Karoubi, M.: Homology of the infinite orthogonal and symplectic groups over algebraically closed fields. Inventiones Math. **73** (1983), 247–250

[Mil]   Milnor, J.: Algebraic K-Theory and bilinear forms. Inventiones Math. **9** (1970), 318–344

[Ser]   Serre, J.-P.: Cohomologie Galoisienne. Lecture Notes in Mathematics **5**. Springer-verlag

[Ser I]   Serre, J.-P.: Local Fields. Grad. Texts in Math. **67**. Springer- Verlag

[Ser II]   Serre, J.-P.: L'invariant de Witt de la forme $Tr(x^2)$. Comm. Math. Helv. **59** (1984), 651–676

[Ser III] Serre, J.-P.: A course in arithmetic. Grad. Texts in Math. **7**. Springer-Verlag

[Sn 1]   Snaith, V.P.: Topological methods in Galois representation theory. C.M. Soc. Monographs, Wiley (1989)

[Sn 2]   Snaith, V.P.: A construction of the Deligne-Langlands local root numbers for orthogonal Galois representations. Topology (2) **27** (1988), 119–127

[Su]   Suslin, A.A.: On the K-theory of algebraically closed fields. Inventiones Math. **73** (1983), 243–249